The Focus and Leverage Improvement Book

Locating and Eliminating the Constraining Factor of Your Lean Six Sigma Initiative

The Focus and Leverage Improvement Book

Locating and Eliminating the Constraining Factor of Your Lean Six Sigma Initiative

By
Bob Sproull

A PRODUCTIVITY PRESS BOOK

First edition published in 2019
by Routledge/Productivity Press
711 Third Avenue New York, NY 10017, USA
2 Park Square, Milton Park, Abingdon, Oxon OX14 4RN, UK

Printed on acid-free paper

International Standard Book Number-13: 978-0-367-00111-7 (Hardback)
International Standard Book Number-13: 978-0-429-44445-6 (eBook)

Library of Congress Cataloging-in-Publication Data

Library of Congress Cataloging-in-Publication Data
Names: Sproull, Robert, author.
Title: The focus and leverage improvement book : locating and eliminating the constraining factor of your lean six sigma initiative / Bob Sproull.
Description: 1 Edition. | New York, NY : Taylor & Francis, [2019] | Includes bibliographical references and index.
Identifiers: LCCN 2018026362 (print) | LCCN 2018021253 (ebook) | ISBN 9780367001117 (hardback : alk. paper) | ISBN 9780429444456 (e-Book)
Subjects: LCSH: Lean manufacturing. | Six sigma (Quality control standard)
Classification: LCC TS155 .S743 2019 (ebook) | LCC TS155 (print) | DDC 658.5/62--dc23
LC record available at https://lccn.loc.gov/2018026362

Visit the Taylor & Francis Web site at
http://www.taylorandfrancis.com

As we go through life, humans continue learning new things. Sometimes we learn through reading books, sometimes it's through personal experiences and sometimes it's through someone we've met along the way. I can honestly say that I have experienced all three of these learning modes. But having said this, I can say without reservation that much of my learning came about when I met Bruce Nelson. For those of you who've never met Bruce, he's a one–of-a-kind guy.

I met Bruce when I worked at the US Army Base, Fort Rucker, Alabama. We were working on improving the maintenance processes on Army helicopters. Bruce worked for another division of the same company, but came to our facility to help us with our improvement efforts. From the very first day I met Bruce, my learnings on the system's approach accelerated upward and have continued since then. Bruce is just a cut above everyone else I have ever met. His method for explaining new concepts is unlike anyone else I have ever met. He has a gift for making complex subjects simple to understand. Bruce introduced me to systems thinking and Theory of Constraints (TOC) Logical Thinking Processes.

I feel honored to have met Bruce Nelson, to have him co-author two books with me and, most importantly, to have him be my friend. I am dedicating this book to Bruce in recognition of everything he has taught me over the years. I can only say, "Thank you Bruce, for all you have done for me and for being my friend."

Contents

Preface

In 2010, Bruce Nelson and I began writing a blog under the name of *Focus and Leverage*, and since then we have posted well over four hundred blog posts on a variety of different subjects. The response to our posts has been overwhelming. The metric we used to track the receptivity of our posts was the number of page views by week. The number of views started at around one hundred per week, but then it started increasing exponentially until it reached well over three hundred thousand per week. We were both amazed and flabbergasted at the level of response we were getting to our posts. Several of our more avid readers recommended that we take what we had written in our blog posts and create a book, summarizing our posts. We thought long and hard about this suggestion and have decided to press forward and have this idea come to fruition.

You will notice that I use the term "we" to explain the concepts presented throughout this book, even though I am the only author listed. It's because Bruce has elected not to be a co-author this time, due to time constraints. But even though he's not listed as a co-author, his writings cover much of this book and make it much more interesting and informative. Much of what we have written in our posts has to do with creating a forum of sorts. That forum is one on continuous improvement methodologies, principles and best practices. After all, continuous improvement has been the hallmark of both our entire careers, and those who know us well know that we love helping others with their efforts.

The layout of this book will be a series of our most popular blog posts, which will be combined to form the chapters in this book. As you can see in the table of contents, unlike many books, one chapter does not necessarily build on the previous one. For example, one of the most popular posts has been our method for teaching people the basics of the Theory of Constraints. We will be discussing our first experiences with TOC, and in so doing, we think you will see why we have embraced it so much. Another topic of discussion will be on our integrated Theory of Constraints, Lean and Six Sigma. We have christened this approach the Ultimate Improvement Cycle, which is also known by the abbreviation TLS. Because these two subjects are similar in nature, they will be linked into the same chapter.

We will also be writing about problems and what we believe is the best approach for solving them. We'll introduce various problem-solving tools and techniques we have used throughout our careers, including tools like Pareto Charts, Causal Chains, Cause and Effect Diagrams, Why-Why Diagrams and a host of other tools we have used in our careers to solve both simple and complex problems. This chapter will also include a session on paths of variation, which you might find to be an interesting read.

In addition to these time-tested problem-solving tools and techniques, we will also be presenting the TOC problem-solving toolset, known as the Logical Thinking Processes (LTPs). These tools were introduced by Dr. Eliyahu Goldratt and have been somewhat modified over the years. We will illustrate the LTPs by use of a case study on how to use TOC Thinking Processes. Bruce has earned the title of a certified Jonah's Jonah, which means he can teach and certify others. And as you will see, he knows and understands the LTPs as well as anyone I know.

Closely related to the LTPs is a technique known as the Goal Tree, which, if followed to its logical conclusion, will teach you how to assess your company and in a single day will help your company create a strategic improvement plan. Often, we have seen students come away from a training session on the LTPs not knowing how to use them. For these people, the Goal Tree has proven its worth many times over.

We will also be writing about two distinctly different methods to manage projects, namely the Critical Path Method (CPM) and Critical Chain Project Management (CCPM). In this chapter, we will discuss project management failure modes, project management negative behaviors that must be overcome, and how best to track projects. We will demonstrate why CCPM is the project management method of choice and how, by using CCPM, your projects can be finished on time, on scope and on budget over 95 percent of the time.

We will also introduce you to the TOC Parts Replenishment Model and will compare it to the traditional Min/Max system. We will demonstrate how, by using the TOC Parts Replenishment Model, you will be able to reduce your facility's parts inventory levels by 40 percent to 50 percent or more, while reducing your stock-outs to nearly zero.

Throughout this book, we will be presenting case studies which were based mainly on work that both Bruce and I have been a part of in our careers. For example, in one of our chapters, we will be presenting case studies on Aviation Maintenance, Repair and Overhaul (MRO) as well as a Hospital Emergency Department setting. You're probably thinking that we may have lost our minds thinking that Healthcare and MRO should be discussed in the same chapter, as though they were related in some way. But as you'll see, they are not so dissimilar after all.

We will also be discussing our technique on how to achieve a motivated work force to actively participate in improvement initiatives. We will be discussing something we refer to as *active listening*. In presenting this technique, we will describe how it has never failed us and how usually it left the leadership team scratching their heads, wondering why such a simple technique has eluded them for all these years.

Another subject that is very important for any organization is which performance metrics they should be using to measure and track their performance. Performance metrics stimulate organizational behaviors, so it's important that an organization selects metrics that result in behaviors that benefit the organization. Closely related to selecting the right performance metrics is the type of accounting system being used. Traditionally, Cost Accounting (CA) has been the system of choice for most organizations. We will discuss in detail another brand of accounting known as Throughput Accounting (TA), which will dramatically influence your organization's profitability. We will demonstrate that the key to organizational profitability is not through how much money you can save, but rather through how much money you can make. And the two approaches are dramatically different!

There will be other discussion topics added as needed, but the ultimate objective will be to share the information that has guided us throughout our careers. We value knowledge and experience sharing above all else, and it is our hope that as you read this book and see all that it has to offer, you will share your learning experience with others. But equally important, it is our hope that you will learn and apply these subjects to improve your company's products and services. Good luck, but as we have written in our other books, luck is Laboring Under Correct Knowledge. You make your own LUCK!

1

Improvement Efforts

INTRODUCTION

How's your improvement effort working for you? If you're like many companies, you've invested lots of money on improvement training, but you're not seeing enough of your investment hitting the bottom line. Like any other investment, you expected a fast and acceptable return on investment (ROI), but it just isn't happening, or at least not fast enough to suit you or your board of directors. Maybe your investment was in Six Sigma, and you've trained hundreds of people to become Yellow, Green and Black Belts, and maybe even a few Master Black Belts? Maybe you've invested a large sum of money training people on Lean Manufacturing and have several Lean Senseis? Or maybe you've gone the Lean Six Sigma route? So why aren't you seeing an acceptable return on investment? You know improvements are happening, because you've seen all of the improvement reports. But you're just not seeing the return on investment that you expected, or at least hoped for.

I too have experienced this dilemma early in and throughout my career. So, early in my career, I performed an analysis of both failed and successful improvement initiatives. What I found changed my approach to improvement forever. What I discovered was that it was all about *focus and leverage.* By knowing where to focus my improvement efforts, I was transformed. In doing so, I discovered something called the *Theory of Constraints* (TOC). TOC teaches us that within a company there are *leverage points* that truly control the rate of money generated by a company. Sometimes these leverage points are physical bottlenecks, but often they are company policies that prevent companies from realizing their true profit potential. In this chapter (and others), we're going to demonstrate what TOC is and exactly how TOC can work for you. We're going to show you how to use the power of TOC, to truly jump-start your improvement efforts. Better yet, we're going to help you turn all of those training dollars (or whatever currency you use) into immediate profits, and then show you how to sustain your efforts over the long haul.

Before we get into the solution, let's take a look at the problem of why your bottom line isn't improving fast enough to suit you or your leadership. If you're like many companies, there seems to be a rush to run out and start improvement projects without really considering the bottom-line impact of the projects selected. Some companies even develop a performance metric that measures the number of on-going projects and attempt to drive the number higher and higher. Instead of developing a strategically focused and manageable plan, many companies, in effect, try to "solve world hunger." They do this instead of focusing on the areas of greatest payback. Many Lean initiatives attempt to drive waste out of the entire value chain, while Six Sigma initiatives attempt to do the same thing with variation. There's nothing wrong with either of these improvement methodologies, but the real benefits occur only if they are focused on the right area.

The real problem with failed Lean and Six Sigma initiatives is really two-fold, namely too many projects and focusing on cost reduction. Many companies simply have too many on-going projects that drain valuable resources needed for the day-to-day issues facing them. Knowing what to do next can really be confusing to managers who have reached their saturation point and are not able to distinguish which projects are vital or important and which ones are not. The economic reality that supersedes and overrides everything else is that companies have always wanted the most improvement for the smallest amount of investment. Attacking all processes and problems simultaneously, as part of an enterprise-wide Lean Six Sigma (LSS) initiative, quite simply overloads the organization and does not deliver an acceptable ROI. In fact, according to the Lean Enterprise Institute annual surveys, the failure rates of LSS initiatives are hovering around 50 percent. With failure rates this high, is it any wonder why companies scale their efforts back, or even abandon the initiatives altogether and back-slide to their old ways?

Earlier we said that focusing improvement initiatives on cost reduction was one of the reasons that many Lean Six Sigma initiatives are failing. Across-the-board cost cutting initiatives are pretty much standard for many businesses. Companies spend inordinate amounts of money on external consultants and in-house training programs, and then focus on ways to reduce costs. Focusing solely on cost reduction is an absolute mistake! So, if this misguided focus isn't right, then what is the right approach? We'll demonstrate why this focus is misguided, but more importantly, where the right focus should be.

Based upon my experiences in a variety of different organizations and industries, the disappointing results coming from Lean and Six Sigma are directly linked to failing to adequately answer three basic questions: What to change? What to change to? and How to cause the change to happen? Take a look at your own company. Are your projects focused on cost reduction? Do you have an army of Yellow, Green and Black Belts? Do you have so many projects that they are bogging down your company? Are your Six Sigma projects typically taking three to six months to complete? Are they providing you with real bottom-line impact, or are they a mirage? We'll begin to address how we can take advantage of the LSS training that's already been provided, to accelerate your company's profits and then demonstrate where to focus your improvement efforts.

The best way we have found to help people understand just what a constraint (i.e. your leverage point) is, and how it impacts the flow or throughput through a process, is by using a simple piping system. These are the actual slides I use to teach the basic concepts of the Theory of Constraints.

Figures 1.1 and 1.2 are drawings of a cross section of pipes, used to transport water. That is, water enters into Section A, then flows into Section B and continues downward until it reaches the receptacle at the bottom of this piping system. I then explain to the audience that it has been determined that more water is needed (i.e. a faster flow rate) and that

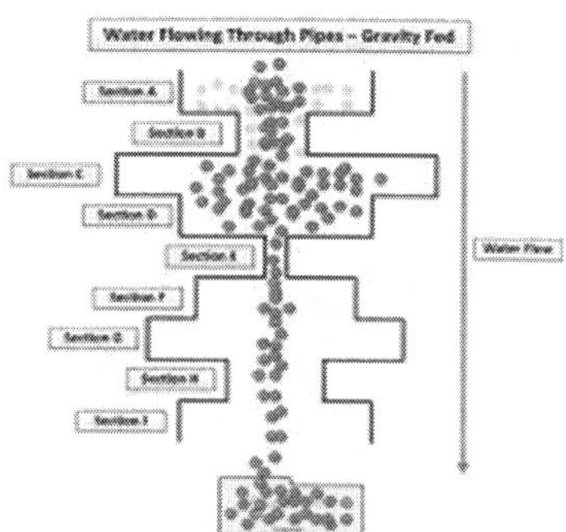

FIGURE 1.1
Piping System Introduction.

Identifying and Exploiting the System Constraint

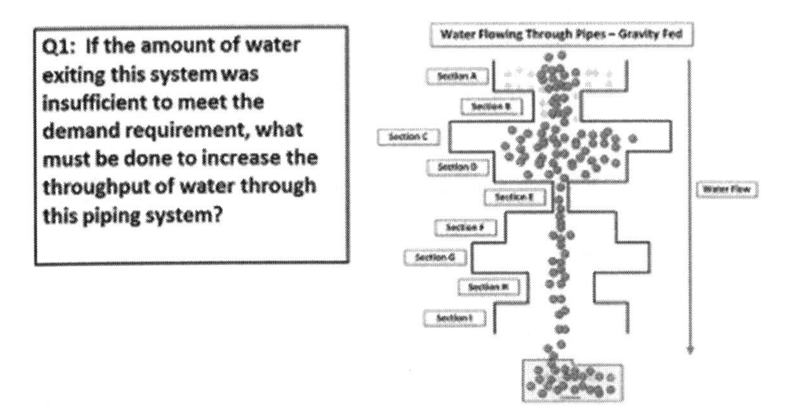

FIGURE 1.2
Actions to Increase Water Flow Rate.

they have been chosen to fix this system. I emphasize that this system is fed via gravity, so they can't simply increase the water pressure. I then pose the question, if enough water isn't flowing through this system, what must be done to make more water flow at a faster rate? Someone in the group will invariably state that to have more water flowing faster through the system, the diameter of Section E must be increased.

I then ask everyone if they understand why they must increase Section E's diameter, and most will answer that they do. For anyone who doesn't, we simply ask the person who first answered our question to explain why this must be done. They usually tell the audience that because of the constricted nature of Section E, water flow is limited at this point. So, the only way to have more water is to expand the size of the pipe in Section E. Since they all now have an understanding of this basic concept, we then move to the next slide.

This next slide, Figure 1.3, reinforces what we just explained, but then I ask another important question about how large the new diameter should be. In other words, what would the new diameter of Section E depend upon? The key point here is that this is supposed to demonstrate that demand requirements play a role in determining the level of improvement needed to satisfy the new demand requirements. In other words, "How much more water is needed?"

In Figure 1.4, we demonstrate the new diameter of Section E and how water is now flowing at a much faster rate than before the diameter change.

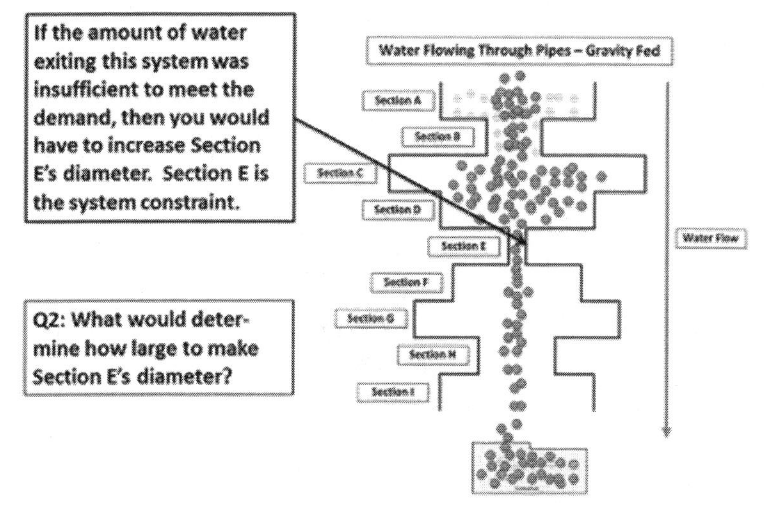

FIGURE 1.3
Identifying the System Constraint.

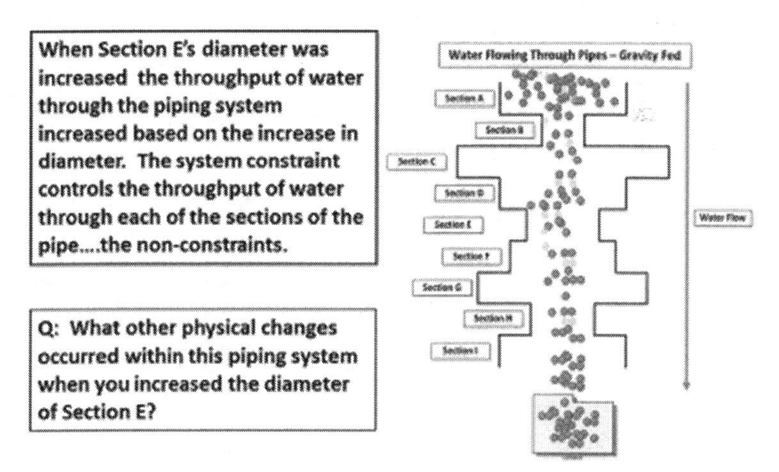

FIGURE 1.4
Exploiting the System Constraint.

The important point I emphasize is that the system constraint controls the throughput of water through every section of pipe, and if we don't subordinate the rest of the system to the same throughput rate as the constraint, there will automatically be a work-in-process (WIP) build-up in front of the constraint.

I then ask the class to identify any other physical changes that have occurred as a result of our exploitation of the constraint (i.e. increasing the diameter of Section E). I give them time to answer this question, and most of the time the group will answer correctly, stating that water is now backing up in front of Section B. I then post the next slide (Figure 1.5) to reinforce what changes to the system have occurred.

Here I point out that, first and foremost, the system constraint has moved from Section E to Section B. I then explain that the new throughput of water is now governed by the rate that Section B will permit. I point to the queue of water, stacked up in front of Section B. I now make the point that if the amount of water is still not enough, then you must decide how to exploit the new system constraint and that the process of on-going improvement is continuous.

In our next slide (Figure 1.6) I ask the question, "Would increasing the diameter of any of the other sections (i.e. non-constraints) have resulted in any more throughput of water through this system?" This question is intended to demonstrate that, since the system constraint controls the throughput of a system, focusing improvement anywhere else in the system is usually wasted effort.

I finish with a Before and After slide (Figure 1.7), just to reinforce how things have changed by focusing on the constraint.

Identifying and Exploiting the System Constraint

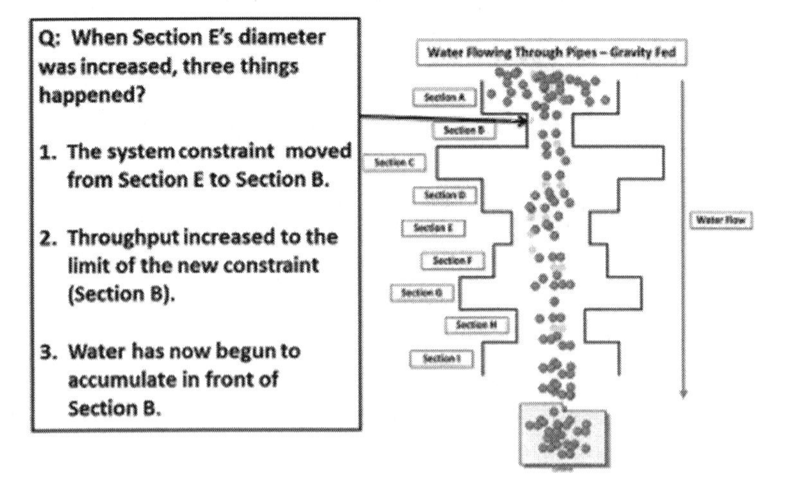

FIGURE 1.5
System Constraint Has Moved.

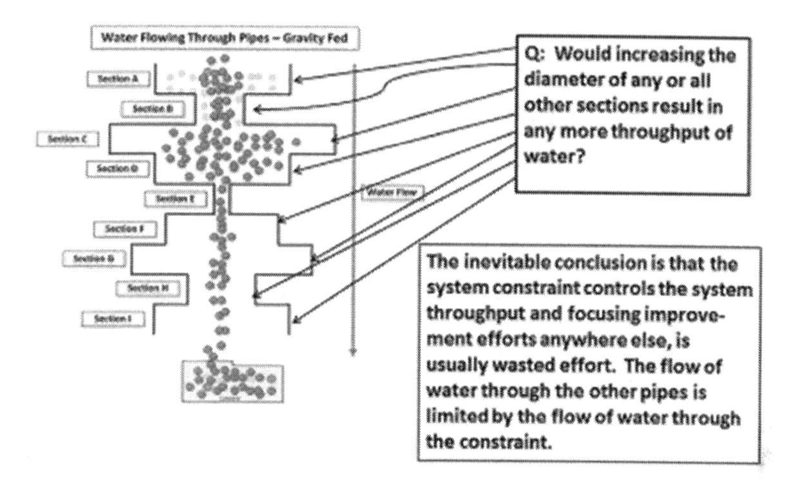

FIGURE 1.6
System Constraint Controls the System Throughput.

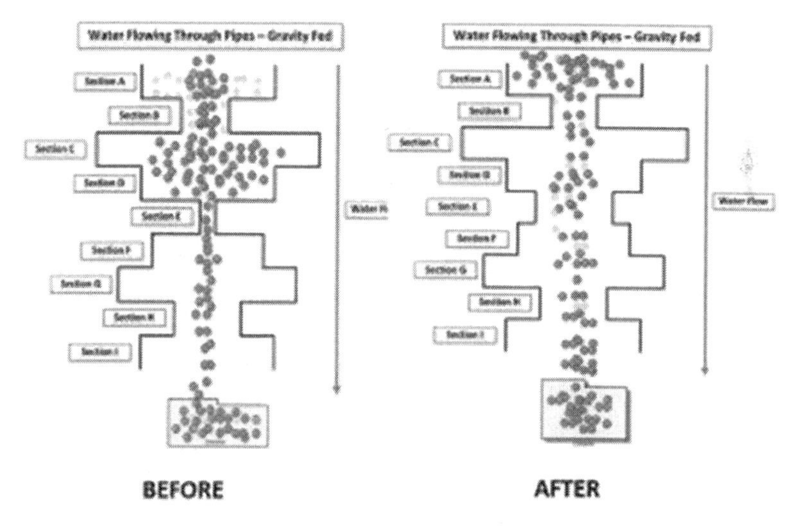

FIGURE 1.7
Before and After.

The important lesson to have learned thus far is that in this simple piping diagram with different diameter pipes, the smallest diameter controlled the throughput of water through the system. I now turn our attention to a simple four-step process used to make some kind of product.

But for anyone new to the Theory of Constraints, I first discuss what the originator of TOC, Eli Goldratt [1], describes as his 5 Focusing Steps:

1. Identify the system constraint.
2. Decide how to exploit the system constraint.
3. Subordinate everything else to the system constraint.
4. If necessary, elevate the system constraint.
5. Return to Step 1, but don't let inertia create a new system constraint.

I do this because I want the class to get the connection from the piping system to the real world. The next slide (Figure 1.8) is the aforementioned simple four-step process, with cycle times for each step listed. I then ask our students to tell me which step is constraining Throughput of this manufactured part. It's been my experience that only about 60 percent of the class makes the connection between the flow of water through the pipes and the flow of product through this process. What I have found to be very effective is to select someone who does understand the connection and have them explain their reasoning. It's important that we don't move on until everyone understands this connection.

The next slide (Figure 1.9) is used to reinforce what their fellow classmates or team members have just explained. I also relate Step 3 of this process to Section E of the piping diagram. That is, I want the class to understand that Step 3 is identical to Section E of the piping diagram, in that both of them are the physical constraints within each of the two systems. I can't emphasize enough just how important understanding this connection really is.

In our next slide (Figure 1.10), I have the class become consultants. They are told that the company who owns this process needs more Throughput.

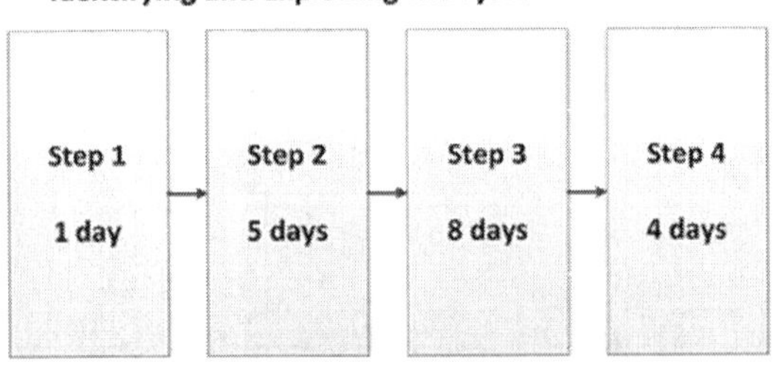

FIGURE 1.8
Where is the System Constraint.

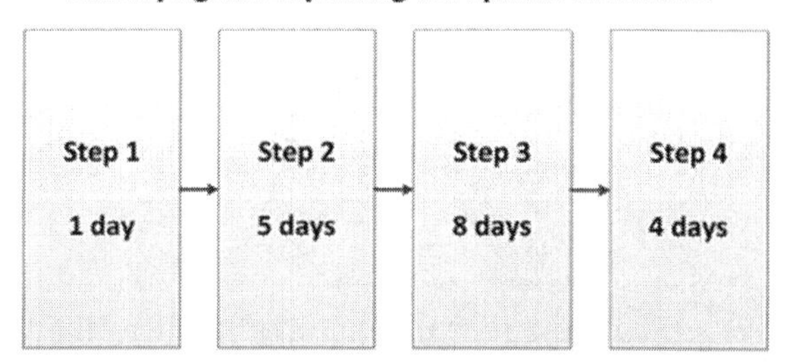

FIGURE 1.9
Identifying the System Constraint.

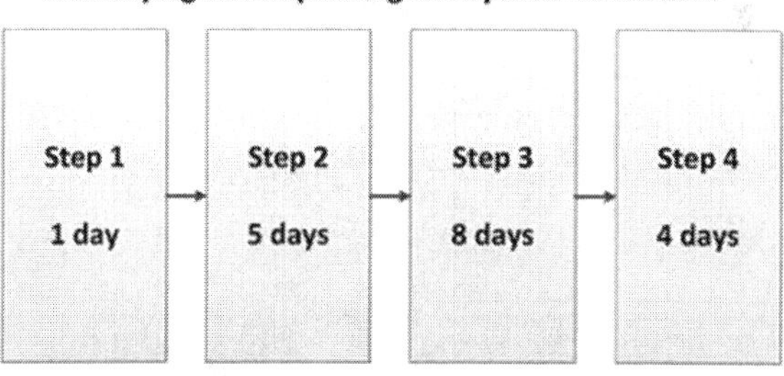

FIGURE 1.10
How do You Increase Throughput.

We ask them what they would do and ask them to explain their answers. I usually break the class up into teams and let them discuss this question as a group, and that seems to work well.

After the team(s) have explained their plan to improve Throughput, I then show them this next slide (Figure 1.11), to reinforce each team's answer on what they would do to increase Throughput.

Because I want the class to understand the negative implications of running each step of this process at maximum capacity, I then ask the class what would happen to the WIP levels if they did run each step to maximum capacity (Figure 1.12).

In the next slide (Figure 1.13), I demonstrate the impact of trying to maximize the performance metrics, *efficiency* or *utilization*, in each step in the process. The key point here is that the only place where maximizing efficiency makes any sense is in the system constraint. The

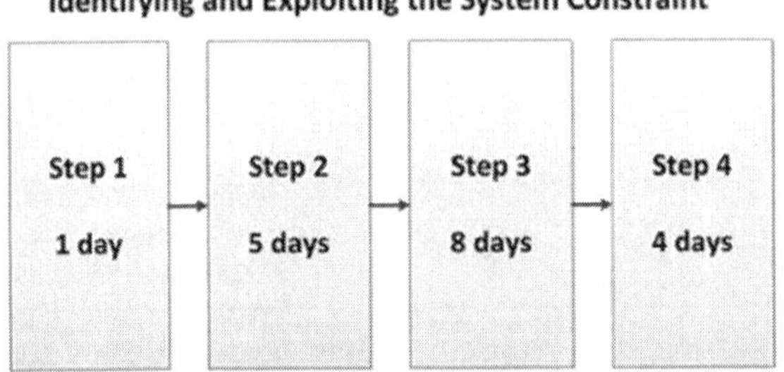

FIGURE 1.11
The Only Way.

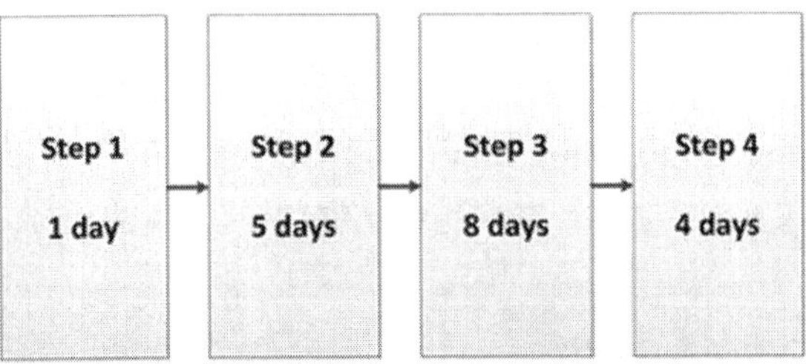

FIGURE 1.12
Balanced Flow and Effect on WIP.

excessive WIP build-up encumbers the process and extends the cycle time of the process, which typically results in a deterioration of on-time deliveries.

I then ask the class, "How fast should each step in this process be running to prevent this excessive build-up of WIP?" This is intended to demonstrate Goldratt's third step, *Subordination*. That is, why it's so important to subordinate every other part of the process to the constraint. This next slide (Figure 1.14) explains in more detail the concept of subordination. Steps 1 and 2 must not be permitted to outpace the constraint, but must also assure that the constraint is never starved. This slide usually creates an epiphany of sorts for the team or class.

I then explain that if the output of the constraint is still below requirements, then we must *Elevate* the constraint, which simply means that we may have to spend some money to increase the capacity of the

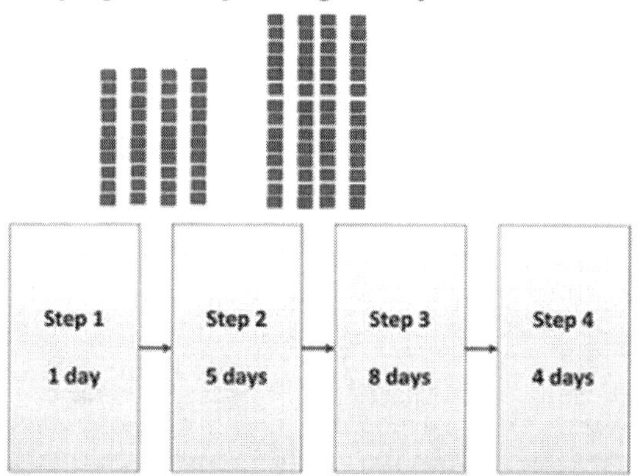

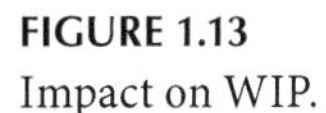

FIGURE 1.13
Impact on WIP.

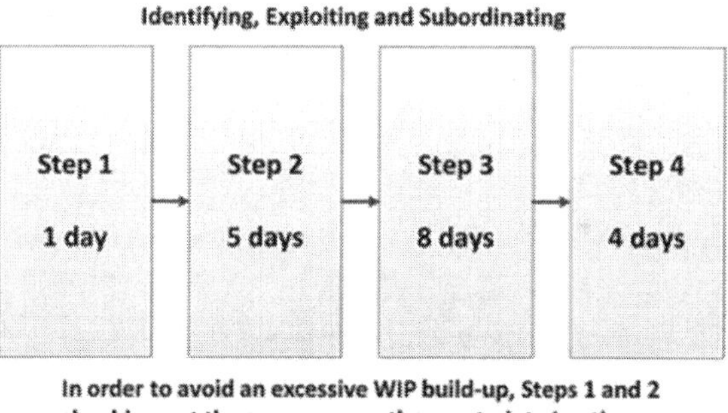

FIGURE 1.14
Avoiding WIP Build-up.

constraint. This can be done by purchasing another machine, using overtime, hiring additional manpower, etc.

My final slide (Figure 1.15) is one that lists Goldratt's 5 Focusing Steps. We talk through each step and relate both the piping diagram and the four-step process to each of Goldratt's five steps.

I have been using this simple method of teaching the concept of the constraint for quite a few years now, and it has worked quite well for me. We strongly suggest that you try it yourself.

1. **Identify the System's Constraints.**
    - What in the system is limiting throughput?
2. **Decide how to Exploit the System's Constraints.**
    - Get the most out of the constraining element without additional investment.
3. **Subordinate/Synchronize the Rest of the System to the Constraint.**
    - Set and implement rules to maximize the capacity of the constraint even if this reduces the efficiency of non-constraint resources.
4. **Elevate the System's Constraints.**
    - Physically increase the capacity of the constraint through acquisition of more resources.
5. **If in the previous steps a Constraint has been broken, go back to Step 1.**

FIGURE 1.15
Goldratt's 5 Focusing Steps.

So, why are we so worried about improving Throughput? Shortly, we will explain why focusing your improvement efforts on increasing Throughput, rather than the typical cost reduction, is the best way to make more money. If cost reduction isn't the best way to make more money, then what is? To demonstrate this to you, I need to introduce you to a different brand of accounting called Throughput Accounting (TA).

TA was developed by Dr. Eli Goldratt, the creator of the Theory of Constraints, to simplify financial decisions in real time. I know this will seem like a lot of new terms, but as you will see, they really are simple or at least much simpler than a traditional Cost Accounting (CA) report. If you can learn these simple terms now, and apply them in your company, you will be well on your way to making more money for your company than you ever dreamed possible.

A NEW TYPE OF ACCOUNTING

To judge whether an organization is moving toward its goal of making more money, three basic questions must be answered:

1. How much money is generated by your company?
2. How much money is invested by your company?
3. How much money do you have to spend to operate your company?

Traditional Cost Accounting, as many of you have experienced, is not only difficult to understand, but it's all about what you did last month or even last quarter. Dr. Goldratt recognized the need to use real-time financial data and created his own version of accounting. He developed the following three simple financial measurements:

Throughput (T)

The rate at which the system generates "new" money, primarily through sales of its products or services. T represents all money coming into a company, minus what it pays to suppliers and vendors. The actual equation for T is

$$T = P - \text{TVC}$$

where P is the price/unit of product or service and TVC is the total variable cost associated with the sale of the product or service. TVC is typically the cost of raw materials, but could also include things like sales commissions, etc., things that vary for each unit of product sold.

Investment or Inventory (I)

The money the system invests in items it plans to sell. Although it includes the price of equipment, buildings, etc., primarily this includes the cost of raw material, WIP and finished goods (FG) inventory.

Operating Expense (OE)

The money spent on turning investment (Inventory (I)) into T, including labor costs, supplies, overheads, etc., basically any expense that is not in the TVC category.

The important point here is that Throughput is not considered T until new money enters the company, by producing and shipping product to its customers and receiving payment. Anything produced that is not shipped and sold is simply Inventory (I), which costs the company money and ties up needed cash. Accordingly, Goldratt defined Net Profit (NP) and ROI as follows

$$\text{NP} = T - \text{OE}$$

and

$$\mathrm{ROI} = (T - \mathrm{OE}) \div I$$

So, with these three simple measurements, T, I and OE, Goldratt reasoned that organizations are able to determine the impact of their actions and decisions on the company's bottom line NP and ROI, *in real time.*

Ok, so why do we say that focusing on T is so important? If you consider inventory reduction, it is a one-time improvement that frees up cash, but after the initial reduction, there's really nothing left to harvest. OE has a functional or practical lower limit, and once it is reached, nothing more can be harvested without having a negative impact on the total organization. That is, companies that typically focus on reducing OE many times engage in layoffs, which if cut too deep will negatively impact the performance of the company, and the morale of the remaining workforce will deteriorate.

What about T? Theoretically, T has no upper limit! As long as you have the sales and you can reduce the cycle time of the constraint, T continues to increase. Yes, T has a practical upper limit, but the potential is there for companies to continue growing profits.

Figure 1.16 represents a graphic of what we just explained. Look at the potential of all three profit components. OE, the target of most improvement projects, is actually the smallest component, followed by Inventory (I), which is slightly higher, but is limited at zero. T, on the other hand, is much larger than either I or OE. The point here is that if you want to drive profitability higher and higher, you must focus your improvement efforts on the process that is limiting your ability to do so. By focusing on the constraint, it's now possible to drive Throughput higher and higher. As you have just seen, TA is much simpler to understand than traditional CA, and it lets you make faster decisions, using real-time data, with decisions linked directly to your company's bottom line.

If TA can be used to make real-time decisions, is it obvious to you that the best decisions are those that result in increasing T, while decreasing OE and I? Keep in mind, we're not suggesting that traditional Cost Accounting should be abandoned, because it is required by law. But, if you're trying to determine where to focus your improvement efforts in real time, then you should be using TA. As explained earlier, when we introduced Goldratt's 5 Focusing Steps of TOC, the Theory of Constraints Process of On-Going Improvement (POOGI) actually works quite well. With all this in mind, we have set the stage for future discussions on why integrating TOC with

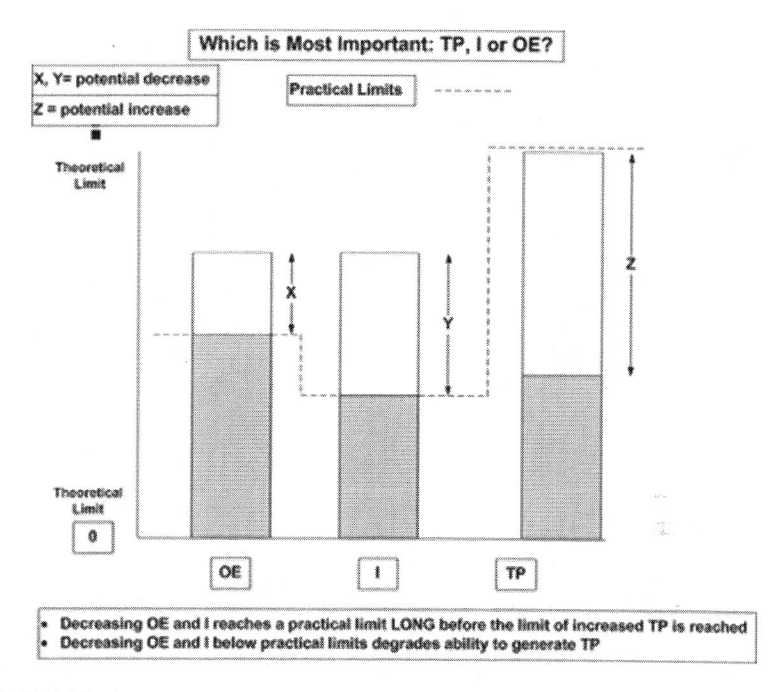

FIGURE 1.16
Order of Importance in Throughput Accounting.

Lean and Six Sigma is the best strategy of all for improvement in your company and supply chain.

We want to back-track now and talk more about the Theory of Constraints 5 Focusing Steps, and how TOC will help you select the right area to focus on to maximize your profits. These five steps form the framework for significant and sustainable improvement in your company. To review, these five steps form our POOGI:

1. Identify the system constraint.
2. Decide how to exploit the constraint.
3. Subordinate everything else to the above decision.
4. If necessary, elevate the constraint.
5. Return to Step 1, but don't let inertia become the constraint.

Let's look at each of these in more detail, so that they make sense to you from an improvement perspective.

Step 1: Identify the system constraint: Since the goal of a typical "for-profit" company is to make more money now and in the future, and since we just learned that focusing on Throughput is the best way to make more money, then we need to identify the *resource* or *policy* in our system that is preventing our company from producing and shipping more goods or providing more services. It is important to understand that the constraint is not always physical. In fact, the majority of the time the constraint is a policy or procedure that limits your company's ability to improve your throughput. In a future chapter, we will discuss the different types of constraints that companies face on a routine basis.

Step 2: Decide how to exploit the system constraint: The key here is to make sure that the constraint's time is never wasted doing things that it shouldn't be doing, and is never starved for work. Remember, every minute lost at the constraint is a missed opportunity to increase throughput. Focus everything you do on reducing the cycle time of the constraint and improving the flow of product (or service) through your system constraint.

Step 3: Subordinate everything else to the previous decision: This step is typically the most difficult of all for most companies to accomplish. By definition, if a process step is not a constraint, it is faster than a constraint and therefore has more capacity to produce than a constraint does. But why would you want a non-constraint to process or produce product at a faster rate than a constraint? If you did, you would only serve to increase WIP inventory within the system, which needlessly ties up cash and extends lead times. Subordinating essentially means to produce at the same rate as the constraint, nothing more and nothing less.

Step 4: If necessary, elevate the system constraint: One of the terms everyone is familiar with is takt time, which is essentially the throughput rate required to meet customer requirements. But what if after all the improvements you've made in Steps 1–3, you still aren't supplying enough product to meet market requirements? Unlike the first three steps, in this step you may have to spend some money by either purchasing a new piece of equipment or even hiring additional labor. Whatever it takes to meet the market demand. This is what elevating the constraint means. Doing whatever it takes to meet the demands of the market.

Step 5: Return to Step 1, but don't let inertia become the constraint: After completing Steps 1–4, you should see significant reductions in the cycle time of the constraint, so you must prepare for a new one to occur. That is, after you essentially "break" your current constraint, another will immediately appear to take its place. This new constraint will require the same improvement cycle that you just went through, and the cycle of improvement continues. In this step, Goldratt said "but don't let inertia become the constraint." What does that mean? Quite simply, during the first four steps of this process, you may have developed and implemented policies or procedures to accommodate your constraint. It is highly likely that these policies and procedures may no longer apply, so you may need to get rid of them. The key here is to keep the improvement process going.

OK, so that's our Process of On-Going Improvement introduced by Goldratt back in the 80s. Think back to our previous discussion on the piping system and the four-step process, and try to imagine your own current POOGI. Does this new version of POOGI apply to your company? We're going to get into the nuts and bolts of how to integrate Lean, Six Sigma and the Theory of Constraints, which is what we call the *Ultimate Improvement Cycle* (UIC). In the next chapter, we will explore the UIC in much more detail.

REFERENCE

1. Goldratt, Eliyahu M., *The Goal—A Process of Ongoing Improvement*, The North River Press Publishing Corporation, Great Barrington, MA, 1984.

2

The Ultimate Improvement Cycle

INTRODUCTION

In Chapter 1, I explained the basic concept of the Theory of Constraints (TOC) Process of On-Going Improvement (POOGI), using Goldratt's 5 Focusing Steps. In this chapter, we want to discuss how best to combine Lean (L), Six Sigma (SS) and TOC to achieve breakthrough profits and Return on Investment (ROI).

The major difference between Lean, Six Sigma and TOC improvement initiatives is simply a matter of *focus and leverage*. While L and SS, or their combined LSS, implement improvements and measure reductions in inventory (I) and operating expense (OE), as well as increases in throughput (T), TOC focuses up-front on T and looks for ways to achieve higher and higher levels. The only way to increase T is to focus on the operation that is limiting it, the *constraint*. We then use Lean to reduce waste and Six Sigma to reduce and control variation, but we do so mainly in the constraint (or on the operation(s) feeding the constraint). Let's look at all three improvement initiatives as if they were stand-alone.

In Figure 2.1, we see the classic Lean improvement cycle. You begin by defining what value is throughout the entire, defined value chain. You then make value flow without interruption, pull to customer demand and then relentlessly pursue perfection. There is no question that a Lean implementation will result in an improved process, but will it result in substantial and rapid bottom-line improvement?

Now let's look at the Six Sigma improvement cycle in Figure 2.2.

In Step 1 you define problems and customer requirements, and then set goals. In Step 2, you measure/collect data to validate and further refine problems. In Step 3, you must analyze the collected data, to develop and validate causal hypotheses. In Step 4, you should improve your processes

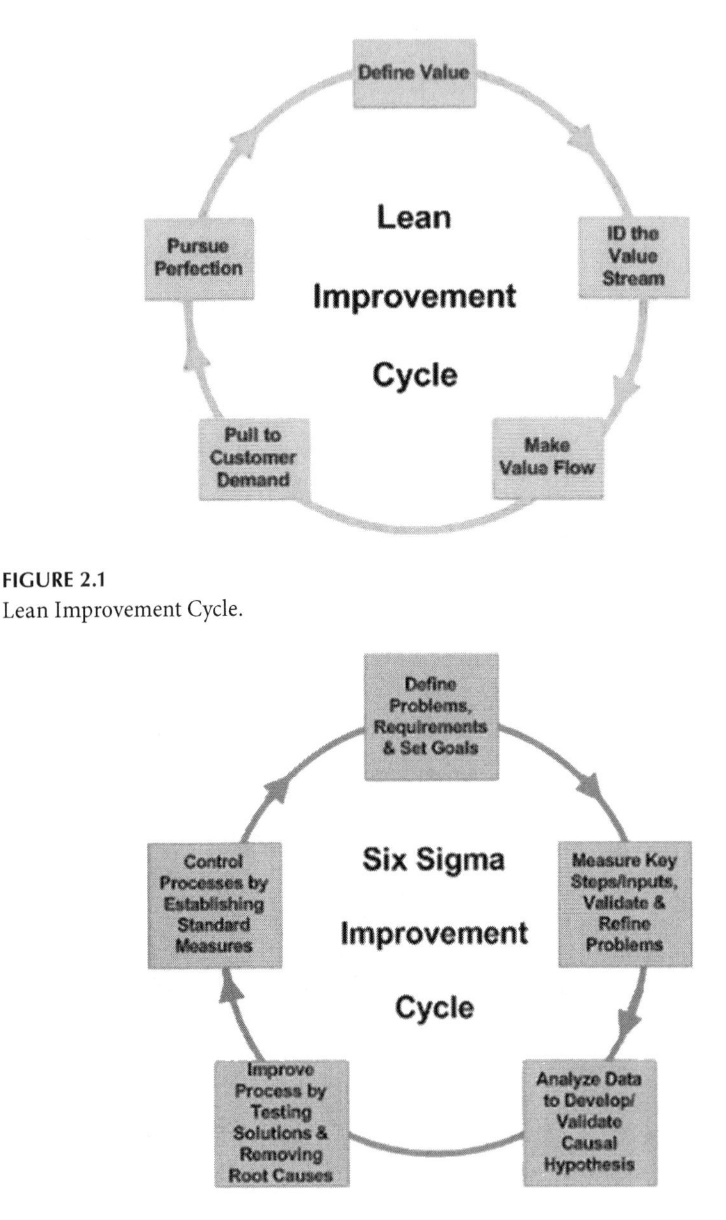

FIGURE 2.1
Lean Improvement Cycle.

FIGURE 2.2
Six Sigma Improvement Cycle.

by developing, testing and implementing solutions and removing root causes. Finally, in Step 5, you must establish and implement standard control measures, which completes the now classic Define, Measure, Analyze, Improve and Control (DMAIC) cycle of improvement. There is no question that Six Sigma creates a much better process, but one of the issues with a typical Six Sigma project is that it takes too long to complete. And have you impacted the bottom line significantly and quickly?

Finally, let's look at the TOC improvement cycle that we discussed in Chapter 1. The next figure (Figure 2.3) lists Goldratt's five improvement steps, Identify, Exploit, Subordinate, and Elevate. The cyclic arrow indicates that for Step 5, after the completion of the Elevate step, you should return to Step 1. In my opinion, the TOC cycle of improvement requires the use of improvement tools and techniques that exist within both Lean and Six Sigma. But having said that, both Lean and Six Sigma need the focusing power of TOC to be effective. The three of them form a symbiosis or interdependence of sorts, that will guarantee improvement of the system in question.

So, if you were to combine the best of all three improvement initiatives into a single improvement process, what might this amalgamation look like? Logic would tell you that you would have an improvement process that reduces waste to make value flow (i.e. through Lean) and reduces and controls variation (i.e. through Six Sigma), but the improvement effort would be focused (i.e. through TOC) on the operation that is constraining throughput. So, think about what this improvement methodology might look like, and I'll show you my version of this integration and then discuss how it all works.

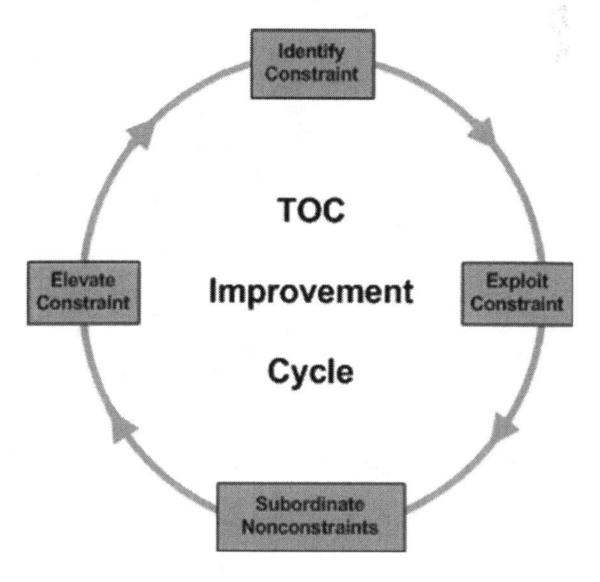

FIGURE 2.3
TOC Improvement Cycle.

Figure 2.4 is a visual of what this integrated methodology looks like, so let's now begin walking through how it all works. This integration weaves together the DNA of Lean and Six Sigma, with the focusing power of TOC, to deliver a powerful and compelling improvement methodology. All of the strategies, principles, tools, techniques and methods contained within all three methodologies are synergistically blended and time-released to yield improvements that far exceed those obtained from doing these three initiatives in isolation from each other.

The *Ultimate Improvement Cycle* (UIC) [1] is not simply a collection of tools and techniques, but rather a viable manufacturing strategy that focuses resources on the area that will generate the highest return on investment. The UIC is all about focusing on and leveraging the operation or policy that is constraining the organization and keeping it from realizing its full financial improvement potential.

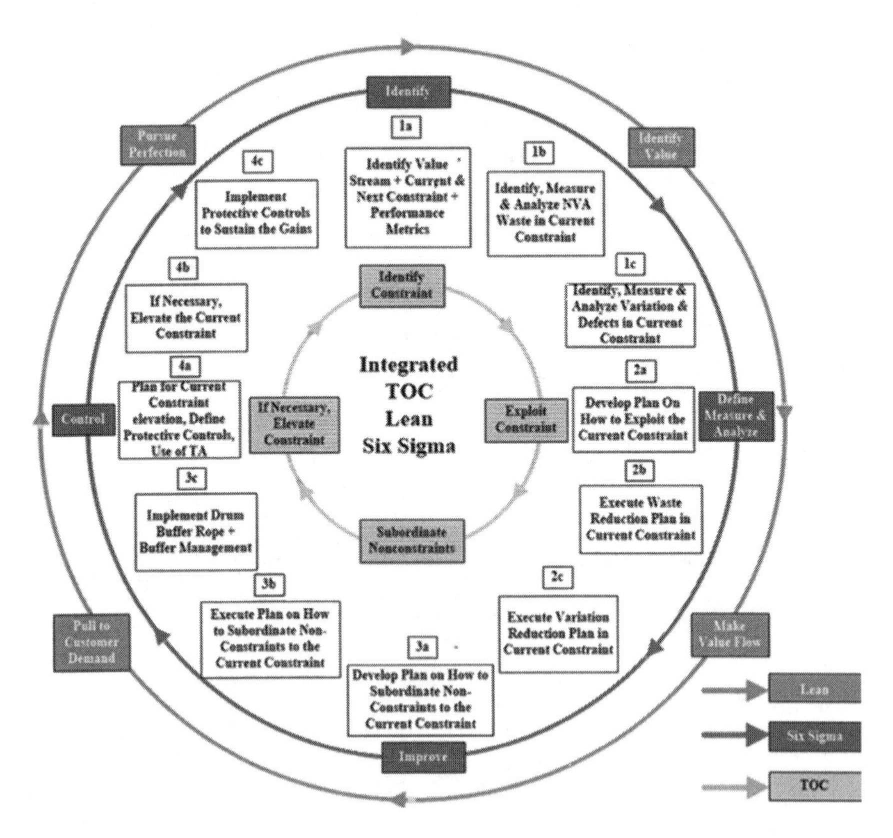

FIGURE 2.4
Integrated TOC, Lean and Six Sigma Cycle.

In Figure 2.4, in Step 1a, we start by identifying the value stream and the current and next constraint and developing appropriate performance metrics to enable us to measure the impact of our improvement efforts. Why is it that we recommend identifying the current and "next" constraint? We do so because as soon as we have improved the current constraint, a new one will immediately appear. So, by identifying the next most logical location of the "new" constraint, we will be prepared to begin improving it immediately (i.e. apply Goldratt's 5 Focusing Steps).

In Step 1b, we identify, measure and analyze non-value-added waste in the current constraint. In Step 1c, we identify, measure and analyze both variation and defects, again in the current constraint. We must also say that, if excessive variation and/or defects are present in the operation feeding or exiting the constraint, then this must be addressed as well. The reason is that materials feeding the constraint must always be available so that the constraint is never starved for work. Likewise, if the process downstream from the constraint is plagued with excessive variation and/or defects, then we risk missing our shipments of finished product to our customers. In Step 2a, we use the information collected in 1b and 1c to develop our plan to exploit the current constraint. After the plan is developed, we then execute the waste and variation reduction plan we developed in Step 2a.

In Step 3a, we then develop our plan on how best to subordinate our non-constraints to the current constraint. In Step 3b we execute the subordination plan developed in Step 3a. In Step 3c, we will implement TOC's scheduling system, known as Drum Buffer Rope (DBR) along with Buffer Management. These two methods when implemented together will guarantee that over-production does not happen. Remember from Chapter 1, one of the primary considerations is that we don't want to create excessive work-in-process (WIP) in our processes, so every effort must be made to only produce what is needed to satisfy orders. Otherwise, lead times will become extended, and there will be shipping delays.

Sometimes, even our best efforts won't free up enough capacity on our constraint to satisfy demand requirements. To counter this possibility, we move on to Step 4a where we develop a plan to elevate our current constraint, as well as defining protective controls, so as not to lose the improvements we have already made. In the Elevation step, we will probably need to spend money, but we must spend it wisely. We should use Throughput Accounting (TA) to help make the right decisions. In Steps 4b and 4c, we execute the plan developed in Step 4a, assuming we need to elevate the constraint. We finish Step 4c by implementing our protective controls to sustain the gains we've made. When we have finished with Step 4c, we must start the rotation again because the constraint will have moved.

In Figure 2.5, we have laid out the improvement tools, actions and focus for our improvement initiative. These are the tools to use, and the actions you will use, at each step of the UIC. As you can see, there are no new or exotic tools being introduced. Instead, in creating the UIC, one of our objectives was to keep things simple. Please keep in mind that there are other tools available that can be used to drive this improvement engine. Our point here was to list some of the more common ones.

In Step 1a, I recommend using either a Value Stream Map (VSM) or a Current State Process Map to identify your value stream, plus your current and next constraint. I also recommend selecting performance metrics that are based upon Throughput Accounting, plus an analysis of your current

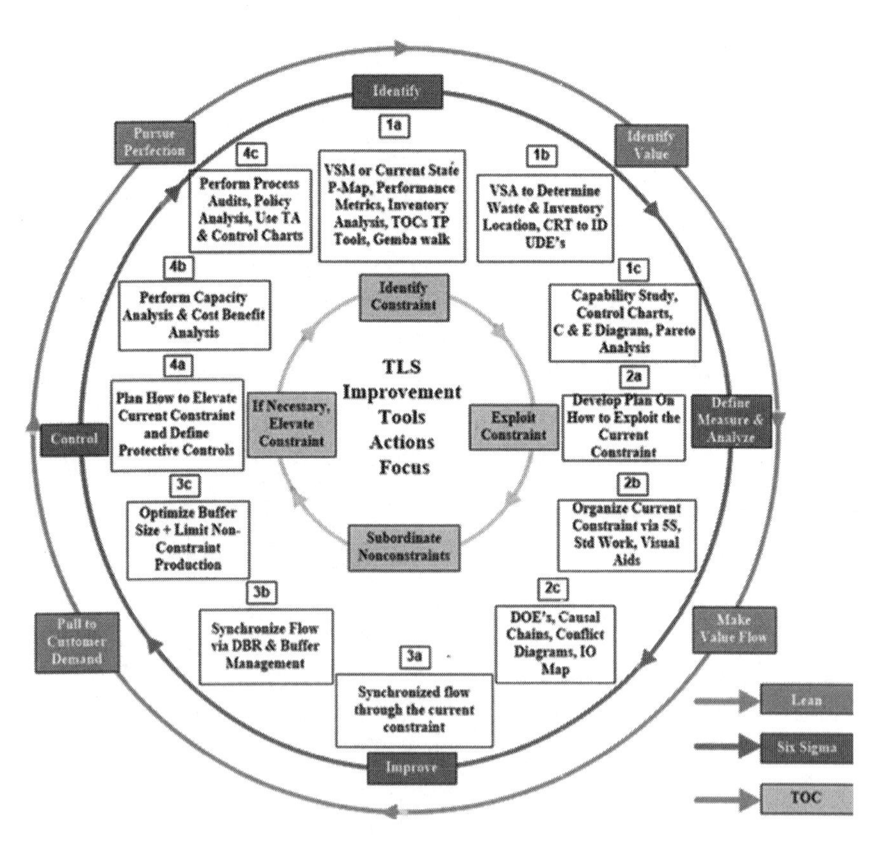

FIGURE 2.5
TLS Improvement Tools, Actions and Focus.

inventory. In addition, I recommend using TOC Thinking Process (TP) tools as well as a Gemba Walk.

In Step 1b, we are performing a Value Stream Analysis (VSA) to determine both locations of waste and excess inventory. My recommendation in this step is to use a Current Reality Tree (to be explained in a later chapter) to identify areas of concern and to identify Undesirable Effects (UDEs) that will impede your performance efforts. In Step 1c, we recommend performing capability studies and implementing control charts where necessary. I also recommend performing a Pareto analysis, and then using tools like a Cause and Effect Diagram to perform a causal analysis. Causal Chains and Why-Why Diagrams are highly effective in this step, as we work to identify cause and effect relationships.

In Step 2a, we will now develop our plan on how best to exploit our current constraint. Some of the tools we will use in Step 2b are organizing the current constraint using 5S (Sort, Set in order, Shine, Standardize, Sustain), standardized work and visual aids. If necessary, in Step 2c, sometimes we will perform a Design of Experiment (DOE), and/or create Causal Chains to identify cause and effect relationships. Sometimes it is necessary to create Conflict Diagrams to identify conflicts. The Conflict Diagram is one of TOC's Thinking Process tools, which we will fully describe in a future chapter. We also recommend using an Intermediate Objectives Map (aka Goal Tree), to further identify key relationships. We plan to dedicate a full chapter to this tool later on in this book.

In Step 3a, we are attempting to create a synchronized flow of product through our current constraint, which is why in Step 3b I recommend using Drum Buffer Rope in combination with Buffer Management to achieve this objective. In Step 3c, we will optimize our buffer size to limit our non-constraint production. In Step 4a, we need to develop our plan on how to elevate our constraint, if it's necessary to do so. In addition, the plan needs to include our selected protective controls. In Steps 4b and 4c, we will execute our elevation plan. I recommend performing a capacity analysis and a cost benefit analysis. In our final step, Step 4c, we need to perform process audits and policy analysis, and implement control charts.

I have added a new element to my original UIC concentric circles (Figure 2.6), first introduced in my book, *The Ultimate Improvement Cycle* [1], with that being your expected TLS deliverables. In Step 1a, you should come away with a complete picture of the system you are attempting to improve in terms of flow, plus the predicted people behaviors and the knowledge that efficiency should only be measured in the constraint.

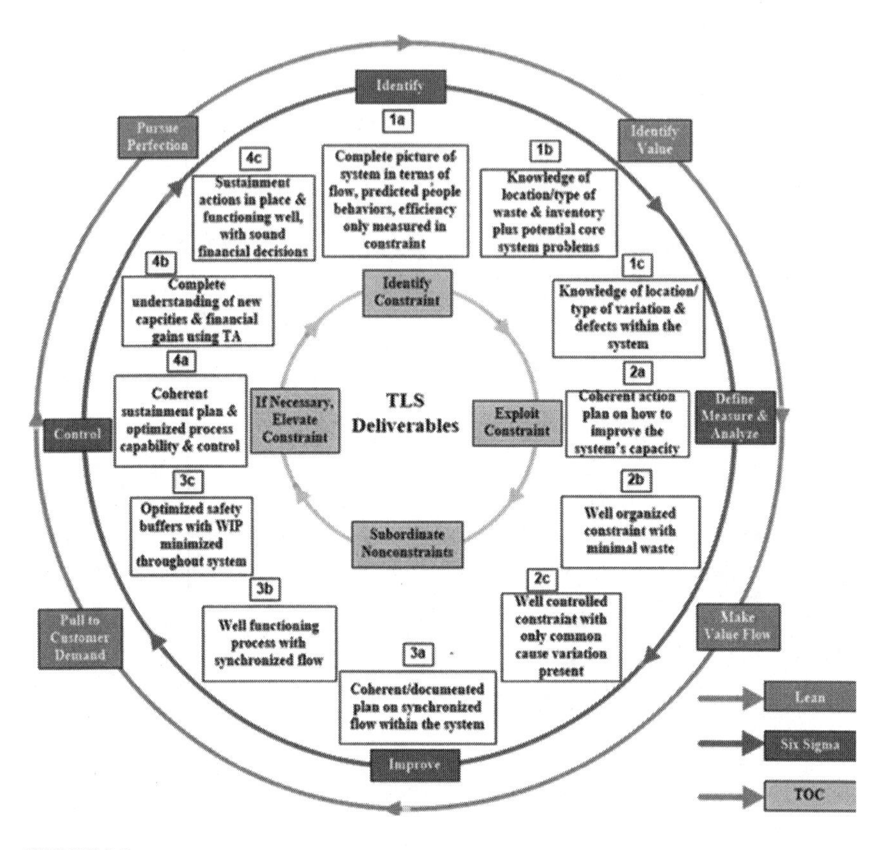

FIGURE 2.6
TLS Deliverables.

In Step 1b, you should come away with the knowledge of location of and type of waste and inventory that exists in your current reality. In addition, you may also come away with a list of potential core system problems. Your take-away for Step 1c should be a working knowledge of both the location and type of variation, plus any recurring problems that currently exist within the system.

Your product from Step 2a is a coherent action plan on how to improve your system's capacity. The product of Steps 2b and 2c is a well-organized and well-controlled constraint, with minimal waste and only controlled, common cause variation present.

In Step 3a, you should come away with a coherent and well-documented plan on how you intend to synchronize flow within your system. Your take-away from implementing your synchronization plan that you developed in Step 3a should be a well-functioning process, exhibiting synchronized flow of products through your system. Your product from Step 3c is an optimized safety buffer, with minimal WIP (thanks to DBR and Buffer Management) throughout your system.

The deliverable from Step 4a is a well-thought-out and clear plan aimed at sustaining the gains you've made that will deliver and sustain an optimized process with excellent process capability and control. From Step 4b, you should have a complete understanding of your new capacities and financial gains from implementing Throughput Accounting. In Step 4c, your sustainment actions will be in place and functioning well, based on sound financial decisions. At this point, your actions should have resulted in a constraint that is no longer your constraint. So, based upon all of your actions and focus, your constraint should now be in a new location. This is the ultimate deliverable, and it's time to return to Step 1a. But this time, your work should be much easier to perform.

The UIC accomplishes five primary objectives that serve as a springboard to maximize revenue and profits as follows:

1. It guarantees that you are focusing on the correct area of the process or system, to maximize throughput and minimize operating expense and inventory.
2. It provides a roadmap for improvement to ensure a systematic, structured and orderly approach to improvement, to maximize the utilization of your improvement resources.
3. It integrates the best of Lean, Six Sigma and TOC strategies to maximize your organization's full improvement potential.
4. It ensures that the necessary up-front planning is completed in advance of changes to the process or organization, so as to avoid the "fire, ready, aim" mindset.
5. It facilitates the synergy and involvement of the entire organization, needed to maximize your full return on investment. In short, you will see a "jump" in profitability!

STEP-BY-STEP EXPLANATION

Step 1a

I will now discuss, in more depth, each step required to achieve these five primary objectives. Step 1 can best be characterized in one word—Identify. As stated earlier, all of Step 1 is a series of activities aimed at identifying, rather than taking action on. What we are attempting to do in this first step is collect information that will become the basis for our well-conceived action plan for improvement. So, as you go through Step 1, we know there will be a near-irresistible urge to make changes, but don't do it yet. The success of the UIC is dependent on the development of a coherent plan and avoiding the "fire, ready, aim" scenario that has become one of the primary causes of failure of many improvement initiatives. Resist this urge!

In Step 1, I have combined identification of the value stream, from the Lean cycle; identification of performance metrics, from Six Sigma; and identification of the current and next constraint from the Theory of Constraints. The flow and inventory analysis is completed by simply reviewing the completed current state VSM or Process Map for location, and volume of inventory within the system. The performance metrics analysis is done by meeting with all departments and leaders to determine what metrics are tracked at all levels of the organization. If your company is like many others, you will be surprised by the number of performance metrics tracked. We also recommend that you determine how the metrics are communicated throughout the organization.

Identifying the current and next constraint is the most important activity in Step 1, simply because the constraint will become the focal point for most of your improvement activities. One of the easiest ways to locate the constraint is by walking the process with your team, during the development of the current state VSM or a Process Map. As you walk this process, you will be identifying both the location and volume of raw material, WIP and finished goods inventory. Typically, the location that has the highest level of inventory will be the current constraint, and the step with the next highest level will be the next constraint, but not always. Look also for policies and procedures that have been implemented that might be policy constraints. An example might be using the performance metrics, manpower efficiency or equipment utilization in every process step. Take your time and do it right, because it will be worth it in the end. Remember, the operative word here is to simply identify, and not take action, just yet.

So, how do you identify the right performance metrics? Performance metrics are intended to serve three very important functions or roles as follows:

1. First and foremost, performance metrics should stimulate the right behaviors.

2. Performance metrics should reinforce and support the overall goals and objectives of the company.
3. The measures should be able to assess, evaluate and provide feedback as to the status of people, departments, products and the total company.

The right behaviors of people and departments are critical to the achievement of the overall goal of the company, but often the metrics chosen encourage and stimulate the opposite behaviors. To be effective, performance metrics must demonstrate the following criteria:

- The metric must be objective, precisely defined and quantifiable.
- The metric must be well within the control of the people or departments being measured.
- The metric must be translatable to everyone within the organization. That is, each operator, supervisor, manager, engineer and operator must understand how his or her actions impact the metric.
- The metric must exist as a hierarchy, so that every level of the organization knows precisely how their work is tied to the goals of the company. For example, if one of the high-level metrics is on-time delivery, then the lower level metric might be cycle time or schedule compliance at individual workstations. Or, if the higher-level metric is parts per million or PPM, then the lower level metric might be the defect rate at an individual workstation.
- The metric should be challenging, yet attainable.
- The metric should lend itself to trend and statistical analysis and, as such, should not be "yes or no" in terms of compliance.

Steps 1b and 1c

In Step 1b, we are attempting to Define, Measure and Analyze (D-M-A) non-value-added (NVA) waste in the constraint operation, while in Step 1b, we are completing the same D-M-A steps, focusing on sources of variation. It is important to remember that in these two steps, we are not taking action to reduce or eliminate waste and variation yet. What I am recommending at this point is to only recognize the existence of waste and variation. It is my belief that one of the primary reasons that many improvement initiatives fail is because of this compulsion to find and react immediately to sources of waste and variation, but I disagree with this approach. In my opinion, it is this compulsion to do everything "right now" that creates a disjointed improvement effort.

Waste and variation reduction efforts are not effective if they aren't done with a systematic plan that ties both of these steps together. You want waste and variation to be attacked simultaneously to ensure that any changes made in the name of waste reduction aren't negatively impacting variation, and vice versa. Remember that for now, because the constraint dictates throughput, and increasing throughput yields the highest potential for significant profitability improvement, you are focusing your waste and variation reduction efforts only on the constraint. The exceptions to this would be upstream process steps causing the constraint to be starved, or downstream process steps scrapping product or causing excessive rework. You cannot ignore these two exceptions. But primarily, you will be focusing your improvement efforts on the constraint.

Figure 2.7 is a tool that I have successfully used many times to search for waste in processes. You will notice that I have listed ten different sources of waste, and symptoms of their existence, instead of the traditional eight. I do this to be as specific as we can in our search. For example, we list overproduction and inventory separately, because the negative impact of overproduction exhibits completely different symptoms from waste of inventory and will require different actions to correct. It helps us focus better.

We now want to turn our attention to variation. There are two types of variability, that you are interested in. No, we're not talking about special cause and common cause. We're talking about processing time variability (PTV) and process and product variability (PPV), which are very different from each other. Sources of PTV are those things that prolong the time required for parts to progress through each of the individual process steps, while PPV are those variables that cause parts' quality characteristics to vary. PPV has a profound impact on PTV, simply because PPV negatively interrupts the process flow. There are many examples of situations that disrupt processes and therefore create variation. Some of the more common examples include unreliable equipment (PTV and PPV), lack of standardized work procedures (PTV and PPV), defective product (PPV and PTV), late deliveries from external and internal suppliers (PTV) and many others.

Variability burdens a factory because it simply leads to congestion, excessive inventory, extended lead times, quality problems and a host of other operational problems. There are two prominent theories on variation and how to treat it. Walter Shewhart's idea was to "minimize variation so that it will be so insignificant, that it does not in any way, affect the performance

Waste Description	Symptoms to Look For
Waste of Transportation	1. Too many forklifts 2. Product has to be moved, stacked and moved again 3. Process steps are far apart
Waste of Waiting	1. Frequent/chronic equipment breakdowns 2. Equipment changeovers taking hours rather than minutes 3. Operators waiting for inspectors to inspect product
Waste of Organization and Space	1. Operators looking for tools, materials, supplies, parts, etc. 2. Large distances between process steps 3. Not able to determine process status in 15 seconds 4. Many different work methods for same process 5. Poor lighting or dirty environment
Waste of Over-processing	1. Rework levels are high 2. Trying to produce perfect quality that isn't required by customer 3. No documented quality standards
Waste of Motion	1. Process steps located as functional islands with no uniform flow 2. Excessive turning, walking, bending, stooping, etc. within the process
Waste of inventory	1. Product being made without orders 2. Obsolete inventory 3. Racks full of product.
Waste of Defective Product	1. Problems never seem to get solved and just keep coming back 2. Independent rework areas have become just another step in the process 3. Excessive repairs
Waste of Overproduction	1. Long production runs of the same part to avoid changeovers and set-up time 2. Pockets of excess inventory around the plant 3. Making excess or products earlier or in greater quantities than the customer wants or needs.
Waste of Underutilization	1. No operator involvement on problem-solving teams 2. No regular stand-up meetings with operators to get new ideas 3. No suggestion system in place to collect improvement ideas 4. Not recording delays and reasons for the delays
Waste of Storage and Handling	1. Many storage racks full of product 2. Damaged parts in inventory 3. Storing product away from the point of use

FIGURE 2.7
Waste Descriptions and Symptoms.

of your product." Taguchi, on the other hand, tells us to "construct (design) the product in such a way, that it will be robust to any type of variation." They're both right, of course. So, what are your options when dealing with the negative effects of variation? There are three ways to handle variation, namely, eliminate it, reduce it or adapt to it. Because it's impossible to totally eliminate variability, you must reduce it as much as possible, and then adapt to the remaining variation. In a later chapter, we will discuss the subject of variation in depth and why it is so important to attack it with a vengeance.

Step 2a

When Goldratt introduced the world to his Theory of Constraints, he did so by laying out his 5 Focusing Steps. His second of five steps was to decide how to exploit the constraint. In other words, how to wring the maximum efficiency out of the constraint. Not just maximizing the efficiency, but because the constraint dictates the performance of the organization, or more specifically dictates the system throughput, how do we maximize our throughput? In Step 2a we will develop our plan on how to exploit the constraint.

My advice to you is very straightforward. If you want your plan to be executed, then keep it simple! Probably many of you have project management software, but I have seen many teams get bogged down in the details of the plan and end up with a failure to launch! Please don't let that happen. Keep it simple, direct and easy to understand, and it will be executed.

Many times, we simply used a Word table or an Excel spreadsheet, because they're easy to use and update. Keep the plan visible and at or near the constraint, since that's where most of the action will be. OK, so what should be in the plan? Figure 2.8 is only a sample of part of a plan I helped to develop for a company, and as you can see, the Constraint Improvement Plan is simple, uncomplicated and straightforward, and follows the actions prescribed in the Ultimate Improvement Cycle. Also, notice that there aren't details on how things like the DOE will be performed, or what will happen during the 5S initiative. This plan is simply intended to be a document that will be used to define the required activities, the expected outcomes and who is responsible for making things happen, and reviewing progress against each of the action items. Each one of the teams will develop their own detailed plan, so again, don't make your Constraint Improvement Plan overly complicated, and be sure to use it for its intended purpose. We have seen so many examples where teams spent an inordinate amount of time on developing the plan at the expense of its execution. Review your plan on a regular basis, and make it visible for everyone to see.

Two final points regarding the improvement plan. The first point is that the order in which you plan and execute is strictly a function of the *current status* of your operation. For example, if you have a major problem with equipment downtime, then activities aimed at reducing downtime should be included in the early stages of your plan. If you have problems related to defective product, then your early efforts should be focused there. The point is, there is no cookie-cutter approach or step-by-step recipe for the order in which activities are planned and executed. It is all dependent upon your own situation and status. In other words, your own current reality.

Action Items	Expected Outcome	Start Date	Complete Date	Responsible	Status	Results
Implement 5S on the constraint operation to reduce NVA time	Reduce NVA time by 25%	6/3/06	6/5/06	B. Kilmer	Complete	NVA time reduced by 31 %.
Develop and implement Drum-Buffer-Rope scheduling system	Improve throughput by 30 % & On-time delivery to >95 %	6/08/06	6/12/06	B. Kilmer	Complete and functioning	Throughput improved by 23 % and on-time delivery improved to 91 %
Develop and implement visual controls and performance metrics in the constraint operation	Improve operation's awareness of process status and progress	6/15/06	6/23/06	B. Kilmer	Complete	Process status is now highly visible
Perform a Gage R & R on measurement system in the constraint.	Total GR&R <10 %	6/3/06	6/6/06	T. Jones	Complete	GR&R = 9.75%
Run a DOE on constraint operation to reduce product variation.	50 % reduction in std deviation	6/6/06	6/13/06	T. Jones	In process	Team formed, study factors & levels identified.
Design and implement manufacturing work cell	Reduce total cycle time by 40 %	6/25/06	6/28/06	B. Kilmer	In process	Kaizen event underway
Reduce changeover time in the constraint operation	Reduce changeover time from 58 minutes to 25 minutes	6/29/06	7/03/08	B. Kilmer	Planning Complete	Kaizen event participants selected
Solve number one defect problem in the constraint operation	Reduce defect level by 50%	6/10/06	6/17/06	T. Jones	Team formed	
Resolve number one cause of machine downtime	Reduce D/T by 50 % & increase OEE by 20%	6/15/06	6/21/06	T. Jones	Team formed	
Improve equipment maintainability	Reduce time to repair by 20 % & OEE by 10%	6/22/06	7/01/06	T. Jones	Team Formed	
Implement pull system in non-constraint operations	Improve Constraint Schedule compliance to >95 %	7/07/06	7/11//06	B. Kilmer	Team selection complete	
Develop and implement process controls and error proofing in the constraint operation.	Reduce product variation by 10 %	6/23/06	6/29/06	T. Jones	Team selected	
Rebalance the process and establish one-piece flow	Improve throughput by 10 %	7/15/06	7/22/06	B. Kilmer	Team Selected	
Develop standardized work instructions in the constraint operation	Reduce processing time by 10 %	6/30/06	7/07/06	T. Jones B. Kilmer	Team selected	

FIGURE 2.8
TLS Action Plan.

The second point to remember is that you must involve the right players as you develop the improvement plan. Believe it or not, the most important members of the team are the hourly operators that will be responsible for operating the new process and making product when the new process is ready. Operators are so often left out of planning activities, when in fact they are the people with the most information—the true process experts. My advice is very clear-cut. If you want your plan to work, then you had better actively involve the operators in its development. And since your operators played a large part in the development of your improvement plan, they will own it and make it happen. In a later chapter, we will discuss what we refer to as "active listening" in more depth.

In addition, the operators must be provided assurance that they are not planning themselves out of a job. The worst possible thing that can happen is that, as cycle times are reduced, or defects and downtime are eliminated, people get moved out of their jobs or worse yet, laid off. If this is your strategy, then I suggest that you stop right now, because it's a strategy for disaster. If this were to happen even one time, you will lose your sense of team and the motivation to improve, so do not lay people off! We realize that business conditions can change, or the economy can take a downturn, and that there are times when you simply can't avoid layoffs. But if people sense that the reason their fellow workers are losing their jobs is because of improvements to the process, then improvements will stop immediately.

NON-CONSTRAINTS

Our improvement focus thus far has been on the constraint, but now it's time to turn our attention to non-constraints. You will recall that the third of Goldratt's 5 Focusing Steps is to *subordinate* everything else to the constraint. Just exactly what is a non-constraint? In TOC jargon, a *constraint* is any resource whose capacity is less than the demand placed on it, and a *non-constraint* is any operation whose capacity is greater than the demand that is placed on it. So, theoretically, constraints limit throughput, while non-constraints do not. But, as you will see, the reality is that this is not always true. So, why did Goldratt believe that it was so important to subordinate everything else to the constraint? To quote Debra Smith [3], "The ability to subordinate will define a company's ability to succeed with the Theory of Constraints. Exploitation of the constraint is dependent upon effective subordination."

The key role of non-constraints is to guarantee that the constraint always has work exactly when it is needed, so as never to allow starvation of the constraint. Constraint starvation translates directly into lost throughput, which negatively impacts profitability. The most effective method I have found to assure that constraint starvation does not occur, is by using a TOC-based scheduling system called Drum Buffer Rope coupled with Buffer Management.

DBR is designed to regulate the flow of product through a production line, based on the processing rate of the most constrained resource, otherwise known as the *Capacity Constrained Resource* (CCR). In a DBR system, the production rate of the CCR is equated to the rhythm of a drum. To protect the drum (CCR) from starvation, a time buffer is placed in front of it, which is the average amount of time required for raw materials to be released into the process and processed by the upstream non-constraints, in time to reach the CCR. To guarantee that product reaches the drum on time, a signaling mechanism, referred to as a *rope*, connects the drum (CCR) to the raw material release for the first operation. Therefore, the first purpose of the rope is to ensure that the CCR is never starved.

By the same token, we want to guard against excess WIP entering the system, and the rope prevents this as well. Incidentally, the derivation of the term DBR is found in Goldratt's book, *The Goal* [1]. So if you haven't ever read it, I strongly encourage you to do so. Because of the importance of DBR, we will now spend some time focusing on the implementation of DBR.

DRUM BUFFER ROPE (DBR)

The first step in any kind of TOC-based implementation is to correctly recognize the constraint, or more specifically, the Capacity Constrained Resource. The slowest resource in any production operation is the CCR, which, in fact, sets the pace for every other part of the process. The danger in out-pacing the constraint will be an increase in both Operating Expense and Inventory. In fact, maximizing production at non-constraints will always result in excessive amounts of WIP, extended cycle times, more labor than is actually required, more need for storage and the need to spend more than is required to secure the needed raw materials. It's important to remember that what we are trying to accomplish is to deliver excellent due-date performance at minimum inventory levels.

The important learning here is that different resources have different capacities. But because of statistical fluctuations and unexpected interruptions, which can never be totally eliminated, your solution must consider these two phenomena. Murphy's expression was, "Anything that can go wrong, will go wrong." We believe that Thompson's Law might also apply. You've probably not heard about Thompson, but his law states that anything that can go wrong already has, you just don't know it.

The cold, hard reality is that the constraint must be protected from "Murphy" (and Thompson) at all times, and DBR does this effectively. DBR utilizes three strategically placed buffers to guard against Murphy (and Thompson) as follows:

- A buffer in front of the constraint to avoid starvation in the event that Murphy (and Thompson) strike any resource in front of the constraint.
- A buffer in front of assembly, if a constraint part is required to complete the assembly.
- A buffer in front of shipping, to assure on-time delivery, in the event that Murphy (and Thompson) strike upstream of shipping.

It is important to understand that these three buffers are generally in the form of time, rather than physical products. The management of these buffers is critical to your success using DBR, so the question becomes, if time is the buffer, how do you know how much time is required? Before we answer this question, consider Figure 2.9, showing what buffer management might look like in a typical production process.

Figure 2.9 is meant to illustrate any of the three buffers previously listed. The basic purpose of these buffers is to provide a signal to everyone involved as to when and when not to require expediting. That is, when a part is late arriving at the buffer, it creates what is referred to as a "hole" in the buffer. Typically, for each type of buffer, there are three zones: a safe zone (green), a caution zone (yellow) and an expedite zone (red). Each zone is equal to one-third of the total calculated time in which the product must arrive at the buffer location. As you might have guessed, if the part arrives at the constraint on time, it is said to be in the green zone time, so there is no cause for alarm. However, if a hole is formed in the yellow or red zones, there is reason to be concerned, and immediate action must be taken to prevent constraint starvation. In the event that parts fail to arrive in the red zone on time, this naturally means that extreme and immediate actions must be taken (i.e. expediting the part). When this happens, the part will definitely arrive late at shipping and

FIGURE 2.9
Buffer Zones.

will ultimately be late arriving at the customer location. I can't emphasize enough that the correct use of these three buffer zones is absolutely critical to DBR's success.

In addition to these buffers, you can also use improvement data for additional improvements. For example, on one hand, if your parts always arrive in the green zone or yellow zone, then there's a good chance that your calculated buffer time is too large and could be reduced. On the other hand, if your parts are consistently arriving in the red zone, then you must increase the size of your buffer times to avoid continually having late orders to your customers. It is also possible that what you have identified as the constraint is not actually the true constraint and that the real constraint lies elsewhere.

In DBR, the drum is the constraining resource, and the pace of the drum sets both the priority and work schedule for your process. The buffer is typically the time schedule in front of the constraint, shipping and any assembly operation needed at the constraint to protect the due dates for all three areas. Your ability to meet these due dates is dependent on achieving two things:

- The buffer must be large enough to handle all of the uncertainties associated with Murphy's (and Thompson's) Law that seem to always occur and eat up buffer time, so that your schedule is achieved.
- The release of raw materials must be done on time, to ensure that the constraint is never starved and that the shipment and customer delivery dates are never compromised.

The rope is simply the length of time needed to complete the processes that are upstream from the drum. It is a signal sent to the beginning of the process, to make sure that raw materials are released on time. So, let's talk about how we calculate these buffers.

Although this may sound overly simplistic, the correct buffer size is only known by monitoring your process. Like I explained earlier, if parts are always arriving in the green and yellow zones, then the buffer is most likely too big. Equally, if the parts are always in the yellow and red zones, then the buffer is usually too small. Goldratt suggested that the initial buffer size can be calculated by taking one-half of the current lead time and dividing that time between the constraint buffer and the shipping buffer. This initial buffer size can then be adjusted up or down, depending on when the parts are arriving. For example, if you find yourself always in the expediting mode, then your buffer is too small, so more time must be added. If you never or rarely experience expediting, then your buffer can be reduced. This attention to buffers is the essence of what we call buffer management.

Each open work order or production batch will have a buffer status that we can calculate. For example, based upon the buffer status we can relate the color code to a percentage of buffer consumed, according to the following guidelines:

- Green Orders result from the buffer status being greater than 67 percent, which means that there is still plenty of time left to complete the order, so no need to expedite.
- Yellow Orders are observed if the buffer status falls between 33 percent and 67 percent, which means that disruptions have taken place somewhere in the process, and there is a high risk that any other disruptions will probably result in a late order. However, there should still enough time to complete the order, without the need to expedite.
- Red Orders happen when the buffer status falls below 33 percent. The inevitable assumption here is that, if there are more disruptions to the order, the order will definitely be late without expediting.

You must always remember that buffer status can change at any time. Because of this, it is recommended that buffer status be checked every shift, and that a First-In-First-Out (FIFO) priority system be followed. This means that the priority order for working on orders should be red, then yellow and then green. There should always be a documented record of why orders might consistently fall into yellow and red zones, because they represent an opportunity for improvement in lead times and on-time delivery rates. These improvements could result in a distinctive competitive advantage for your company.

Figure 2.10, taken from my book *The Ultimate Improvement Cycle: Maximizing Profits Through the Integration of Lean, Six Sigma and the Theory of Constraints* [3], is a graphic of what the DBR system looks like, with the Drum, Buffer and Rope identified appropriately. Smith explains that in addition to the constraint and shipping buffer, there is a third buffer, the assembly buffer. We included this, in the event that your process involves an assembly that requires the part produced at the constraint before it is completed. That must be accounted for as well, if part of your process includes assembly.

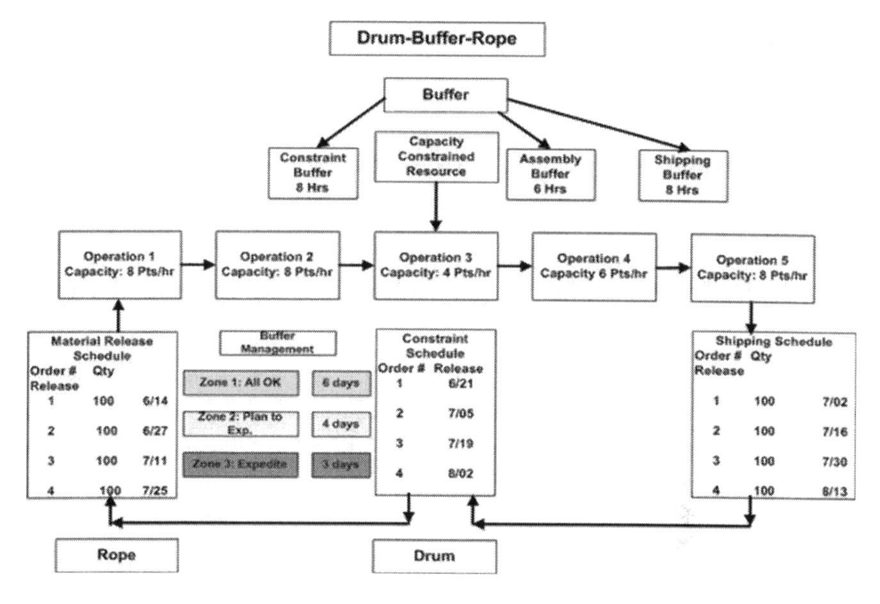

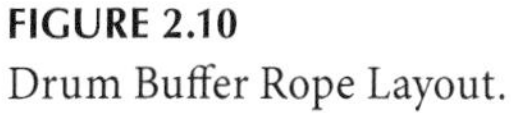

FIGURE 2.10
Drum Buffer Rope Layout.

One of the key points to remember regarding the implementation of Drum Buffer Rope is that it won't be sustainable unless traditional performance metrics like operator efficiency and equipment utilization are abandoned, except at the constraint. So, if these metrics aren't the correct metrics to track in non-constraints, then what metrics should we track? The fact is, non-constraint metrics should be measuring how well they are doing with respect to buffer management. Remember, the key properties of good performance metrics are

- They must be objective, precisely defined and quantifiable.
- They must be well within the control of the people or process that is being measured.
- They must stimulate the right behaviors.

So, what behaviors are we trying to encourage in non-constraint operations? Think for a moment about the function of non-constraints. First, a non-constraint in front of a constraint must never starve the constraint. If this happens, throughput is lost to the system. Conversely, a downstream non-constraint must never be permitted to produce scraps or excessive rework, because that too would be throughput lost to the system. Second, unnecessary inventory in front of the constraint adds needless cost to the organization. Third, the constraint operation sets the pace for all other resources, which means that schedule compliance in a non-constraint is critical. So, based on these three behaviors, there are four metrics that should be considered:

- Workstation availability: The percentage of time the non-constraint was available to make product.
- Yield: The percentage of "good" product produced for the constraint operation (or for downstream operations supplying the shipping buffer).
- On-Time Delivery to the next operation: The percentage of compliance to the DBR schedule.
- Protective Buffer (parts and/or time) in front of the constraint, assembly and shipping. Percentage of protective buffer that is less than needed or too much.

The final decision on which metrics to use for a non-constraint is clearly situation dependent, but the decision should be based on

- Never starving a constraint buffer (or assembly or shipping buffer).
- The negative impact of too little or too much inventory in the system.
- Compliance to schedule.

Let's now continue our discussion on the UIC by looking at Steps 3a, 3b and 3c.

Steps 3a, 3b and 3c

In Step 3a, which is a very important step, we must put together and document a coherent plan on how we are going to synchronize flow through the system. This will include our DBR plus Buffer Management. In Steps 3b and 3c, we must come away with a well-functioning and optimized safety buffer that results in WIP being minimized throughout the total system, to optimize the flow of parts through our processes and system.

Steps 4a, 4b and 4c

In Step 4a, our deliverable is the development of a coherent sustainment plan that optimizes our process capability and control. Goldratt's fourth step tells us that, if necessary, we must elevate the constraint. Increasing the capacity of the constraint can be done in a variety of ways, like using overtime, adding resources, purchasing additional equipment and so on. One thing you must keep in mind when you are elevating the constraint is what will happen when you are successful with this elevation. Remember back in Step 1, we said that you should identify both the current and next constraint? We did this for a reason.

Suppose you have decided that to elevate the constraint, you must purchase a new piece of equipment. Your justification should only demonstrate the throughput improvement up to the limit of the next constraint. That is, if your current constraint is currently producing five parts per hour, and you are purchasing a new machine that will double that to ten parts per hour, then this improvement is only correct if the next slowest resource is at ten parts (or above) per hour. If, for example, the next constraint in the process is only producing seven parts per hour, then you really can only claim a gain in throughput of two parts per hour for the new equipment. All we are saying is that you must make sure you consider the total process when making your decision on how to elevate the current constraint.

What we should end up with from Step 4b is a complete understanding of our new capacities and financial gains from using Throughput Accounting. Finally, in Step 4c, we should have our sustainment actions in place and functioning well, with sound financial decisions being made. The final step, Step 4c, in the UIC, is to Implement controls to sustain the gains. Sustaining the gains is the hallmark of great organizations, so how do we do this?

Of course, if you have chosen the right performance metrics, and you're tracking them religiously, this is one way, but is it the best way? One of the most effective tools to protect and preserve your accomplishments is to use a simple process audit. A typical process audit is a series of questions asked to the line leader and/or supervisor to demonstrate the status of the process. These questions should be focused on how elements of your new process are working and to demonstrate that they are being used as intended.

For example, suppose part of your plan involved implementing a control chart. Your questions would be focused on whether or not it's being used and if out-of-control data points are being investigated and acted upon. You would of course then check the status of the control chart, to verify the responses. These audits should not be planned in advance, but rather be done in a random time frame.

One of the reasons audits of this nature fail to add value is that many times they are announced in advance. Anyone can look good for a day, if they are given enough advance notice. We also advocate having leadership perform some or all of these audits, simply because it adds credibility to them. If leadership thinks that the audits are important, then everyone else will too. In addition, we recommend scoring the audit by percent compliance, and posting the audit score as demonstrated in Figure 2.11.

With any luck, by following the steps we've presented you will have increased the capacity of your constraint to meet the demands of the marketplace. If this is not the case, then other more extreme actions must be taken, which usually involve spending some money. You have eliminated much of the waste and reduced variation, both of which have had a positive impact on cycle time and throughput, but you're not quite there yet as far as producing enough product.

You have just completed one revolution of the Ultimate Improvement Cycle, and things should be greatly improved. Your throughput has increased, your cycle times are reduced, your quality is better, there is less variation and uncertainty, your on-time deliveries are much better and your bottom line is much healthier. Don't stop here! Move on to your next revolution and start the process all over again. This is your new Process of On-Going Improvement … your POOGI. Good luck!

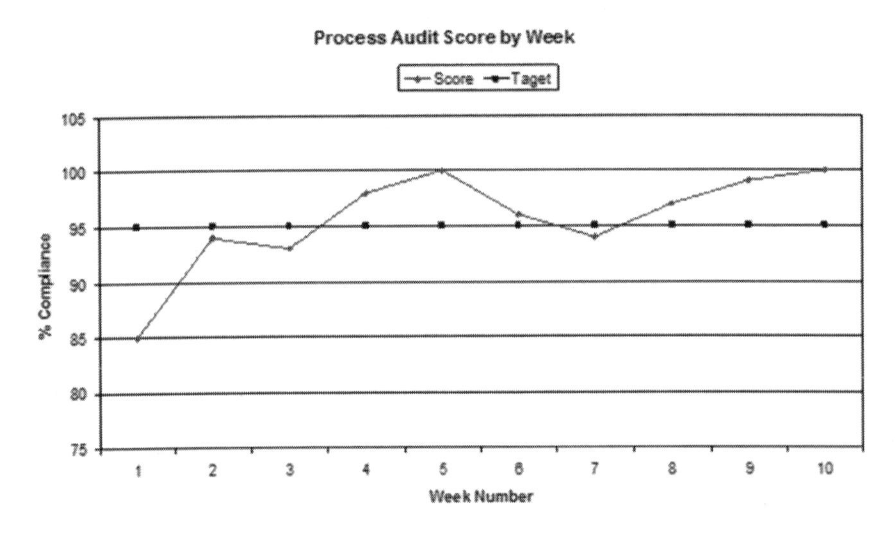

FIGURE 2.11
Process Audit Scoring Graph.

REFERENCES

1. Goldratt, Eliyahu M., *The Goal—A Process of Ongoing Improvement*, The North River Press Publishing Corporation, Great Barrington, MA, 1984.
2. Sproull, Bob, *The Ultimate Improvement Cycle—Maximizing Profits Through the Integration of Lean, Six Sigma and the Theory of Constraints*, CRC Press, Taylor & Francis Group, Boca Raton, FL, 2009.
3. Smith, Debra, *The Measurement Nightmare—How the Theory of Constraints Can Resolve Conflicting Strategies, Policies and Measures*, St. Lucie Press, Boca Raton, FL, 2000.

3

How to Implement the UIC

INTRODUCTION

In our last chapter, we completed the first rotation of the Ultimate Improvement Cycle (UIC), so now it's time to get started with your own cycle of improvement. I have hopefully convinced you of its value for your company. If I have convinced you, then you probably are wondering about the best way to get started. "Do I go out and just start at Step 1a of the Ultimate Improvement Cycle?" The answer is no, because if you did that, you would almost immediately begin hitting barriers and obstacles that would limit your success or maybe even question the validity of this cycle of improvement. So, if not Step 1a, then what?

Let's first consider the question of what we are attempting to do. You need to start by accepting that the basic goal of all "for-profit" organizations is to make money now and in the future. If you're already making money, perhaps your goal might be better stated as, "to make more money now and more money in the future." If this is your goal, then the question you would ask yourself is, "What is preventing me from making more money now and more money in the future?" My experience tells me that there are a host of things that prevent companies from making more money.

In its most basic form, making money involves generating revenue that is greater than what it costs to generate it. Obviously, if operating expenses are too high, and you aren't generating enough revenue, then you won't be making money. So the question is, just how do you generate more revenue? Assume for a moment that you have more orders than you have capacity to fill them. Since you are unable to satisfy market demand, it follows that your throughput is too low. If your throughput is too low, it must also mean that your cycle times are too long. It follows then that the key to generating more revenue must be reducing cycle times. How do we reduce cycle times? Let's first look at something called Little's Law.

Little's Law states that cycle time equals work-in-process (WIP) divided by Throughput (i.e. CT = WIP/T). It should be clear that reducing cycle time implies reducing WIP, as long as Throughput remains constant. So, if you have large amounts of WIP, then clearly you have an opportunity to reduce cycle time. But what if you don't have large amounts of WIP in your plant (I'm betting you do, though)? How else might we reduce cycle times?

We know that cycle time is equal to the sum of all processing times for each process step. We also know that cycle time is the sum of all value-added time, plus all non-value-added time in the total process. So, if we want to decrease cycle time, then we have three choices:

1. We can reduce value-added time.
2. We can reduce non-value-added time.
3. We can do some of both.

As is demonstrated in Figure 3.1, non-value-added time by far and away accounts for the largest percentage of total cycle time in all processes. This would imply that, if we significantly reduce non-value-added time in our process, then we could significantly reduce cycle time, which would, in turn, significantly improve our on-time delivery, Throughput and revenue. So, what are these non-value-added times that we have referred to? Just think about which activities add value, versus those that do not. Let's make a list.

1. Transport time—moving product from point A to point B.
2. Setup time—converting a process from one configuration to another.
3. Queue time—time spent waiting to be processed.
4. Process batch time—time waiting within a batch.
5. Move batch time—time waiting to move a batch to the next operation, which could also include time in storage.
6. Wait-to-match time—time waiting for another component to be ready for assembly.
7. Drying time—time waiting for things like adhesives to become ready to be assembled.
8. Inspection wait time—time waiting for products to be inspected.

Value-Added

Non-Value-Added

FIGURE 3.1
Value Added to Non-Value Added Comparison.

There might be others we could add to our list, but for now, assume this is our list. Which of these items add value? Clearly none of them does, so they would all be classified as non-value-added. There obviously are things we could do to reduce each one of these. For example, process batch time is driven by the process batch size, so we could do two things that would reduce this time. We could optimize the batch size that we produce, and in conjunction with this, we could reduce the time required for setup. In doing these two things, we would probably also reduce the move batch time, and maybe even the wait-to-match time. Clearly, these actions would reduce the overall cycle time.

But even if we were successful in reducing cycle time, we would not realize a single piece of Throughput unless we reduced the processing time and non-value-added time of the operation that is constraining the Throughput—the constraint. Any attempts to reduce processing times in operations that are not constraining Throughput are, quite simply, wasted effort.

The key to making more money now and in the future is tied to two single beliefs, focus and leverage. In Theory of Constraints (TOC) terminology, these two beliefs of focus and leverage are fundamental to the idea of exploiting the constraint. If you want to increase your Throughput, then there is only one effective way to accomplish it. You must leverage the operation that is limiting your Throughput, your constraint operation! And how do you leverage your constraining operation? You do so by focusing your available improvement resources on your constraint and reduce the non-value-added and value-added times within the current cycle time. It's really that simple!

So, are you ready to begin your own cycle of improvement now? Not quite. There are other important things that you must consider before beginning your own cycle of on-going improvement. There are ten prerequisite beliefs that apply to your organization that I believe must be considered before you begin your journey.

TEN PREREQUISITE BELIEFS

We told you not to just jump right into the UIC and begin the improvement process. In this section, we're going to define the ten prerequisite beliefs that your organization must embrace before your organization will be able to successfully implement and navigate through the Ultimate Improvement Cycle:

1. Believing that leveraging the constraint, and focusing your resources on the constraint, is the key to improved profitability. Because of this, the constraint can never sit idle.
2. Believing that it is imperative to subordinate all non-constraints to the constraint. If you violate this key belief, your throughput will not improve, and your WIP will grow to unacceptable levels, thus draining cash from your coffers.
3. Believing that improving your process is a never-ending cycle. You must be ready to re-focus your resources when the constraint moves, and it will move eventually.
4. Believing that involving and empowering your total workforce is critical to success. Your work force has the answers, if you will first listen to what they have to say and then engage them to design the solution.
5. Believing that abandoning outdated performance metrics, like efficiency and utilization, reward or incentive programs and variances is essential to moving forward. As Goldratt says in his book, *The Goal*, "Show me how you measure me, and I'll show you how I'll behave." These outdated metrics and practices are archaic tools from the past, so you must let them go.
6. Believing that excessive waste exists in your process, and that it must be reduced or removed. Studies have confirmed that typical processes have less than 10 percent value-added work, meaning that waste accounts for 90 percent of the available time.
7. Believing that excessive variation is in your process and that it must be reduced and then controlled. One of the keys to growth in profitability is consistent and reliable processes. Processes are full of variation and uncertainty, and unless and until variation is reduced and then controlled, moving forward will be difficult.
8. Believing that problems and conflicts must be addressed and solved. You can no longer afford to hide problems with inventory. When problems arise, you must stop the process and take the time to solve them. By solving them, we're talking about finding and eliminating the root cause(s).
9. Believing that constraints can be internal, external, physical or policy related or any combination of the four. In the real world, over 90

percent of all constraints are policy related. Policies and procedures must be scrutinized, changed and sometimes thrown in the garbage and replaced with policies that make sense.

10. Believing that the organization is a chain of dependent functions and that systems thinking must replace individual thinking. It is no longer acceptable to focus on improving a single step in the process, if it isn't the weakest link. This focus on local optima must be replaced with system optimization.

If your entire operation is ready to accept the prerequisite beliefs of constraint focus and leverage, then you have taken the first step, but it must include everyone and every department. Your entire organization must become focused on the leveraging power of the constraining operation. If you can't do that, then there simply is no need to continue. Unless and until all functional groups within your organization are singing from the same sheet of music, you simply will not make any progress.

Of all the TOC focusing steps, Subordination will be the most difficult one to apply. It simply means that every decision made, and every action taken by the entire organization, must be done so based on its impact on the constraining resource. And when we say the entire organization, we mean everyone! Subordination also means that over-production must not occur, simply because we don't want to clog the system with excess WIP.

Accounting must provide real-time decision-making information to the organization, and not hold onto financial measures that are based on what happened last month or even last quarter. Accounting must also eliminate outdated performance metrics like utilization and efficiency in non-constraint operations, because they mean absolutely nothing.

Purchasing must order parts and materials based on the rate of consumption, especially at the constraint, and stop ordering in large quantities or only on the basis of lowest cost to satisfy another outdated performance metric, purchase price variance.

Sales and Marketing must understand that unless and until the current constraint is broken, they must not make hollow promises on delivery dates to obtain more orders to supplement their sales commissions.

Engineering must respond quickly to the needs of production, to assure timely delivery and updates to specifications. Maintenance must always prioritize their work, based on the needs of the constraining operation, including preventive and reactive maintenance activities. If there is an inspection station that impacts the constraint Throughput, then inspectors (if they exist) must always provide timely and accurate inspections so as to never cause delays that negatively impact the flow of materials into and out of the constraint. Finally, Production Control must stop scheduling the plant on the basis of forecasts that we know are using the outdated algorithms contained within the Manufacturing Resource Planning (MRP) system. DBR should be the scheduling system of choice.

As you identify the constraint, and subordinate the rest of the organization to the constraint, there will be idle time at the non-constraints. If you are like many organizations that use total system efficiency and/or utilization as key performance metrics, then you will see both of them predictably decline. You are normally trying to drive efficiencies and utilizations higher and higher at each of the individual operations, under the mistaken assumption that the total efficiency of the system is the sum of the individual efficiencies. In a TOC environment, the only efficiencies or utilizations that really matter are those measured in the constraint operation. You may even be using work piece incentives in an effort to get your operators to produce more, and I'm sure many of you are using variances as a key performance metric. Efficiencies, utilizations, incentives and variances are all counterproductive!

Believe me, no matter how good you think your processes are, they are full of waste and variation. You must accept the premise that every process contains excessive amounts of both waste and variation that are just waiting to be identified, removed, reduced and controlled. Your job will be to locate, reduce and hopefully eliminate the major sources of both. Variation corrupts a process, rendering it inconsistent and unpredictable. Without consistency and control, you will not be able to plan and deliver products to your customers in the time frame you have promised. Waste drives up both operating expense and inventory, so improvements in both of these will fall directly to the bottom line as you improve the Throughput of your process and, more specifically, your constraining operation. Yes, you will observe waste in your non-constraint operations, but for now focus your resources only on the constraint!

If your organization has truly accepted these ten prerequisite beliefs and all that goes with them, then you are ready to begin this exciting journey that has no destination. But simply saying you believe something can be hollow and empty. It is your day-to-day actions that matter most. Review these ten prerequisite beliefs as a group on a regular basis, and then hold your employees and *yourself* accountable to them. Post them for everyone to see. Utilizing the Ultimate Improvement Cycle, and true acceptance of and employment of these ten prerequisite beliefs, will set the stage for levels of success you never believed were possible!

There has been push-back by some people on the whole concept of Throughput Accounting (TA). As a result, they don't buy into using TA as a reason for integrating TOC, Lean and Six Sigma. So, let's put the financial side of this integration to the side for a moment. In addition to the financial case made for integrating these three improvement methodologies, there are other rational and logical reasons why this integration works so well. In attempting to answer which of these three initiatives a company should use, or "which tune a company should dance to," Thompson [1] presents an excellent summary of the fundamental elements, strengths and weaknesses for each improvement initiative. In doing so, Thompson has inadvertently (or perhaps purposely) answered the underlying question of why the three improvement initiatives should be combined and integrated, rather than choosing one over the other.

The first four columns in Table 3.1 reflect the summary of Thompson's comparison (i.e. the initiative, fundamental elements, strengths and weaknesses). I have added a fifth column, "Counter Balance," that demonstrates how the strengths of one initiative counter balance or compensate for the weaknesses of the others. As a matter of fact, by comparing each of the weaknesses and strengths of each of the three initiatives, we see that all of the weaknesses of each individual initiative are neutralized by one or both of the strengths of the other two. This is such an important point for those companies that have experienced implementation problems for any of the three individual improvement initiatives done solo. Let's look at several examples.

Table 3.1 tells us that Weakness 1 in Lean, "May promote risk taking without reasonable balance to consequence," is counter balanced by Six Sigma Strength 3, "The focus on reduction of variation drives down risk and improves predictability." One thing we know for certain is that as we reduce variation in our process we reduce risk, and our ability to predict future outcomes improves dramatically. This is the cornerstone of statistical process control, which means that risks can be minimized if we rely on this Six Sigma strength to do so. Continuing, Lean Weakness 2 tells us that we may not provide sufficient evidence of business benefit for traditional Cost Accounting. This weakness is countered by both Six Sigma Strength 2, the data gathering provides

TABLE 3.1

Strengths, Weaknesses and Counterbalances

Initiative	Fundamental Elements	Strengths	Weaknesses	Counter Balance
Lean	The cause of poor performance is wasteful activity. Lean is a time-based strategy and uses a narrow definition of waste (non-value-adding work) as any task or activity that does not produce value from the perspective of the end user. Increased competitive advantage comes from assuring every task is focused on rapid transformation of raw materials into finished product.	1. Provides a strategic approach to integrated improvements through Value Stream Mapping and the focus on maximizing the value-adding-to-waste ratio. 2. Directly promotes and advocates radical breakthrough innovation. 3. Emphasis on fast response to obvious opportunities (just go do it). 4. Addresses workplace culture and resistance to change through direct team involvement at all levels of the organization.	1. May promote risk taking without reasonable balance to consequence. 2. May not provide sufficient evidence of business benefit for traditional management accounting. 3. Has a limitation when dealing with complex interactive and recurring problems (uses trial-and-error problem solving).	1. Six Sigma strength # 3 2. Six Sigma strength # 2 and TOC strength # 4 3. Six Sigma strength # 1 and TOC strength # 3.

(Continued)

TABLE 3.1 (CONTINUED)

Strengths, Weaknesses and Counterbalances

Initiative	Fundamental Elements	Strengths	Weaknesses	Counter Balance
Six Sigma	The cause of poor performance is variation in process and product quality. Random variations result in inefficient operations causing dissatisfaction of customer from unreliable products and services. Increased competitive advantage comes from stable and predictable processes allowing increased yields, improving forecasting and reliable product performance.	1. The rigor and discipline of the statistical approach resolves complex problems that cannot be solved by simple intuition or trial and error. 2. The data gathering provides strong business cases to get management support for resources. 3. The focus on reduction of variation drives down risk and improves predictability.	1. Statistical methods are not well suited for analysis of systems integration problems. I can calculate sigma for a product specification, but I am not sure how to establish sigma for process interactions and faults. 2. The heavy reliance on statistical methods by its very nature is reactive, as it requires a repetition of the process to develop trends and confidence levels. 3. The strong focus on stable processes can lead to total risk aversion and may penalize innovative approaches that by their nature will be unstable and variable.	1. Lean strength # 1 and TOC strength # 2 2. Lean strength # 2 and Lean strength # 3 3. Lean strength # 2

(Continued)

TABLE 3.1 (CONTINUED)

Strengths, Weaknesses and Counterbalances

Initiative	Fundamental Elements	Strengths	Weaknesses	Counter Balance
TOC	The cause of poor performance is flawed management technique. Systems logic is used to identify constraints and focus resources on the constraint. The constraint then becomes the management fulcrum.	1. Provides simplified process and resource administration through a narrow focus on the constraint for management of a process as well as improvement efforts (exploitation). 2. Looks across all processes within a systems context to assure that limited resources are not overbuilding non-constraint capability (the local optimization problem). 3. Distinguishes policy versus physical constraints. 4. Provides direction on appropriate simplified measures (throughput, inventory and operating expense).	1. Overemphasizing exploitation of the constraint may lead to acceptance or tolerance of wasteful non-constraint tasks within the process. 2. If the underlying process is fundamentally inadequate no matter how well managed it may not achieve the goals and objectives. 3. Does not directly address the need for cultural change. TOC change process is very technically oriented and fully acknowledges the need for TQM and other improvement methods.	1. Lean strength # 1 and Six Sigma strength # 2 2. Lean strength # 2 3. Lean strength # 4

strong business cases to get management support for resources and by TOC Strength 4, provides direction on appropriate simplified measures (Throughput, Inventory and Operating Expense). As we have stated many times before, traditional Cost Accounting induces us to make incorrect decisions, so by adopting Throughput Accounting practices, from the Theory of Constraints, we will have sufficient evidence to make changes to our process, assuming we are focusing on the constraint operation.

Lean Weakness 3 states that Lean has a limitation when dealing with complex interactive and recurring problems (uses trial and error problem solving) and is countered by Six Sigma Strength 1, the rigor and discipline of the statistical approach resolves complex problems that cannot be solved by simple intuition or trial and error, and TOC Strength 3 (distinguishes policy vs. physical constraints). One of the Six Sigma tools that permit us to solve complex interactive and recurring problems is Designed of Experiments (DOE). DOE identifies significant factors that cause problems, and identifies insignificant factors that do not. TOC Strength 3 helps us in two ways.

First, if the problem we are facing is a policy constraint, we use TOC's Current Reality Tree to identify it, and TOC's Conflict Cloud to solve it. Both of these strengths will compensate for this weakness in Lean.

Now let's look at one of the Six Sigma and TOC weaknesses and see how they are compensated for by other strengths. For example, Six Sigma Weakness 2, the heavy reliance on statistical methods, by its very nature is reactive as it requires a repetition of the process to develop trends and confidence levels. This weakness is off-set by Lean Strength 2, directly promotes radical breakthrough innovation, and by Lean Strength 3, emphasis on fast response to opportunities (just go do it). Likewise, TOC Weakness 3, TOC's inability to address the need for cultural change, is off-set by Lean Strength 4.

In the same way, if we compare all of the weaknesses in Lean, Six Sigma and TOC to the strengths found in the other initiatives, the three initiatives not only complement each other, but they rely on each other. Table 3.1 is from Steven W. Thompson, "Lean, TOC or Six Sigma: Which tune should a company dance to?" from an article in e-newsletter *Lean Directions*. So, in addition to the demonstrated financial benefits of this symbiotic trilogy, we now see evidence from a logical perspective as to why they should be implemented in unison as a single improvement strategy.

TYPES OF CONSTRAINTS

Until now, we have focused on identifying the physical constraints in a manufacturing process, and how to break them. But, what if the constraint isn't located inside the process? What if the constraint is policy related, or a non-physical entity, such as the efficiency metric? Let's take a look at what this means, in terms of our improvement effort.

Bill Dettmer [2] explains that, "Identifying and breaking constraints, becomes a little easier if there is an orderly way to classify them." Dettmer tells us that there are seven basic types of constraints as follows:

1. Resource/Capacity Constraints
2. Market Constraints
3. Material Constraints
4. Supplier/Vendor Constraints
5. Financial Constraints
6. Knowledge/Competence Constraints
7. Policy Constraints

Let's take a look at each of these constraints and what they mean to us for our improvement efforts.

Resource/Capacity Constraints

This type of constraint exists when the ability to produce or deliver the product is less than the demands of the marketplace. That is, the orders exist, but the company has insufficient capacity to deliver. These types of constraints have been what we have been discussing since we started this chapter and why using the UIC will increase capacity (throughput) to high enough levels. In fact, eliminating this type of constraint leads directly to the next one.

Market Constraints

This type of constraint exists when the demand for a product or service is less than the capacity to produce or deliver the product or service. That is, the company has not developed a competitive edge to realize enough orders for their product or service. Market constraints come about simply because the company is unable to differentiate itself from its competition. So, how can a company differentiate itself? Quite simply, there are four primary factors associated with having, or not having, a competitive edge, as follows:

1. **Quality**: In its most basic form, quality is a measure of how well a product conforms to design standards. The secret to becoming quality competitive is first, designing quality into products; second, the complete eradication of special cause variation; and third, developing processes that are both capable and in control.
2. **On-Time Delivery**: This factor requires that you produce products (or deliver services) to the rate at which customers expect them. This means that you must have product flow within your facility that is better than that of your competition. As you now know, this involves identifying, focusing on and improving your constraint. It also involves reducing unnecessary inventory that both lengthens cycle times and hides defects.
3. **Customer Service**: This simply means that you are responsive to the needs of your customer base. Customers must feel comfortable that if their market changes, their supply base will be able to change right along with them, without missing a beat. If the customer has an immediate need for more product, the supplier that can deliver will become the supplier of choice.
4. **Cost**: This factor is perhaps the greatest differentiator of all, especially in a down economy. But having said this, low cost, without the other three factors, will not guarantee you more market share. The good news is, if you are improving throughput at a fast enough rate, the amount you charge a customer for their business can be used to capture it. So, as long as your selling price is greater than your totally variable costs, the net flows to the bottom line.

Material Constraints

This type of constraint occurs because the company is unable to obtain the essential materials in the quantity or quality needed to satisfy the demand of the marketplace. Material constraints are very real for production managers, and over the years they have been such a problem that material replenishment systems like MRP and Systems, Applications and Products (SAP) were developed in an attempt to fix them. However, as you know (or should know), MRP and SAP haven't delivered the needed fix and as a result companies have spent millions of dollars needlessly, because these systems haven't addressed the root cause of the shortages.

Supplier/Vendor Constraints

This type of constraint is closely related to material constraints, but the difference is that suppliers are inconsistent because of excessive lead times in responding to orders. The net effect is that because the raw materials are late arriving, products cannot be built and shipped on time.

Financial Constraints

This type of constraint exists when a company has inadequate cash flow needed to purchase raw materials for future orders. Under this scenario, companies typically must wait to receive payments for an existing order before taking any new orders. An example of this type of constraint is a weak accounts receivable process whereby companies deliver products, but payments take long times to be received and posted.

Knowledge/Competence Constraints

This type of constraint exists because the knowledge or skills required to improve business performance or perform at a higher level are not available within the company. An example of this is a company purchasing robotics but failing to develop the necessary infrastructure and knowledge to support the new technology. What typically happens is the equipment breaks down and remains down for extended periods of time, thus losing needed throughput.

Policy Constraints

Last, but certainly not least, is the policy constraint, which includes all of the written and unwritten policies, rules, laws or business practices that get in the way of moving your company closer to your goal of making more money now and in the future. In fact, Dettmer tells us [3], "In most cases, a policy is most likely behind a constraint from any of the first six categories. For this reason, TOC assigns a very high importance to policy analysis." The most common examples of policy constraints include the use of performance metrics like operator efficiency or machine utilization where there is a push to maximize metrics in all steps in the process, when in reality maximizing them in the constraint is the only place that matters.

In the next chapter, we will discuss something referred to as the Goal Tree, which was originally introduced by Bill Dettmer as an Intermediate Objectives Map (IO Map). Dettmer later changed the name of the IO Map to the Goal Tree.

REFERENCES

1. Thompson, S. W., Lean, TOC or Six Sigma: Which Tune Should a Company Dance To? Society of Manufacturing Engineers, 2005, January 15, 2005.
2. Dettmer, H. William, *Breaking the Constraints to World Class Performance*, Quality Press, Milwaukee, WI, 1998.
3. Dettmer, H. William, *Goldratt's Theory of Constraints: A Systems Approach to Continuous Improvement*, Quality Press, Milwaukee, WI, 1996.

4

The Goal Tree

INTRODUCTION

Many people who have gone through training on the Theory of Constraints (TOC) Thinking Process (TP) tools have come away from the training somewhat overwhelmed and somewhat speechless, to a degree. Some "get it," and some just don't. Let's face it, the TP tools are pretty intimidating, and after receiving the training I have seen many people simply walk away, feeling like they were ill-prepared to apply whatever it is they had supposedly just learned. Even for myself, when I completed my first iteration of this training, I had this same feeling. And in talking with others, there was a general confusion about how to get started. For the average person, the TP tools are just not easy to grasp, so they end up kind of putting them on the back burner, rather than taking a chance on making a mistake using them.

The other complaint I have heard many times is that a full TP analysis typically takes many days to complete, and let's face it, a regular executive team typically doesn't have that kind of time to spend on this activity, or at least they feel like they don't. Well for everyone who feels the same way, or maybe has gone through the same Jonah training as I did and feels somewhat hopeless or confused, I have hope for you. That hope for you is another logic diagram, currently known as the *Goal Tree*. We say currently because the man responsible for creating the Goal Tree, Bill Dettmer [1], originally referred to this tool as an Intermediate Objectives Map (IO Map), but has elected to change its name in recent years. Before going any further, I want to make sure everyone understands that I am a huge proponent of TOC Thinking Processes! But having said that, I am an even bigger fan of the Goal Tree. Why? Because most people grasp what the Goal Tree will do for them right away, and how simple it is to learn, construct and apply. Many of the people I have trained on the Goal Tree have emailed me, telling me they wished they had learned this tool many years ago. They learn it and apply it right away!

Bill Dettmer tells us that his first exposure to IO Maps/Goal Trees was back in 1995 during a management skills workshop conducted by another TOC guru, Oded Cohen, at the Goldratt Institute. In recent years, Dettmer has written much about the IO Map (now referred to as a Goal Tree), and he now uses it as the first step in a full Thinking Process analysis. Bill is passionate about this tool and believes that it defines the standard for goal attainment and its prerequisites in a much more simple and efficient way. I happen to agree with Bill and believe that the Goal Tree is a great focusing tool to better demonstrate why an organization is not meeting its goal. And because of its simplicity, it is not only easy to learn, but also, it's much easier to teach others in your organization how to use it than the full TP analysis.

There are other advantages of learning and using the Goal Tree, including a better integration of the rest of the TP tools that will accelerate the completion of Current Reality Trees (CRTs), Conflict Resolution Diagrams (CRDs) and Future Reality Trees, if you choose to use them. But what I really like about the Goal Tree is that it can be used as a stand-alone tool, resulting in a much faster analysis of the organization's weak points and then a rapid development of an improvement plan for your organization. I have been teaching the Goal Tree for quite a few years and can state unequivocally that the Goal Tree has been the favorite of most of my classes and workshops.

One of the lessons I always encourage my students to follow is that they should always learn a new tool and then make it their own. That message simply means that even though the "inventor" of a tool typically has a specific use in mind, tools should be continually evolving, and such was case for me with the Goal Tree. Personally, I have attempted to transform this tool into one that most people grasp and understand in very short order, and then see its usefulness in a matter of minutes or hours, rather than days.

When using any of TOC's Thinking Process tools, there are two distinctly different types of logic at play, *sufficiency* and *necessity*. Sufficiency-based logic tools use a series of *if-then statements* that connect cause and effect relationships between most of the system's undesirable effects. Necessity-based logic uses the syntax, *in order to have x, I must have y* or multiple *y*s. The Goal Tree falls into the category of necessity-based logic and can be used to develop and lay out your company's strategic and tactical actions that result in successful improvement efforts.

As mentioned earlier, the Goal Tree dates back to at least 1995 when it was casually mentioned during a Management Skills Workshop conducted by Oded Cohen at the A.Y. Goldratt Institute, but it was not part of that workshop, nor did it ever find its way into common usage as part of the Logical Thinking Process (LTP). It was described as "a kind of Prerequisite Tree without any obstacles."

Dettmer tells us that he never thought much about it for the next seven years, until in late 2002 when he began grappling with the use of the LTP for developing and deploying strategy. At that time, Dettmer had been teaching the LTP to a wide variety of clients for more than six years and had been dismayed by the number of students who had substantial difficulty constructing Current Reality Trees and Conflict Resolution Diagrams of sufficient quality. According to Dettmer, they always seemed to take a very long time to build a CRT, and their CRDs were not always what he would characterize as "robust." He claimed they lacked reference to a "should-be" view of the system—what *ought* to be happening. It occurred to Dettmer that the Goal Tree he'd seen in 1995 could be modified and applied to improve the initial quality of CRTs. As time went on, Dettmer began to realize that the Goal Tree could serve a similar purpose with CRDs. In 2007, Dettmer published a book, *The Logical Thinking Process—A Systems Approach to Complex Problem Solving* [2] that introduced the world to this wonderful tool, and I always highly recommend this book.

Dettmer tells us that one of the first things we need to do is define the system boundaries that we are trying to improve, as well as our span of control and sphere of influence. Our span of control means that we have unilateral change authority, while our sphere of influence means that at best, we can only influence change decisions. Dettmer explains that if we don't define our boundaries of the system, we risk "wandering in the wilderness for forty years."

THE GOAL TREE STRUCTURE

The hierarchical structure of the Goal Tree consists of a single Goal and several entities referred to as Critical Success Factors (CSFs). The CSFs must be in place and functioning if we are ever going to achieve our stated goal. The final piece of the Goal Tree are entities referred to as Necessary Conditions (NCs), which must be completed to realize each of the CSFs.

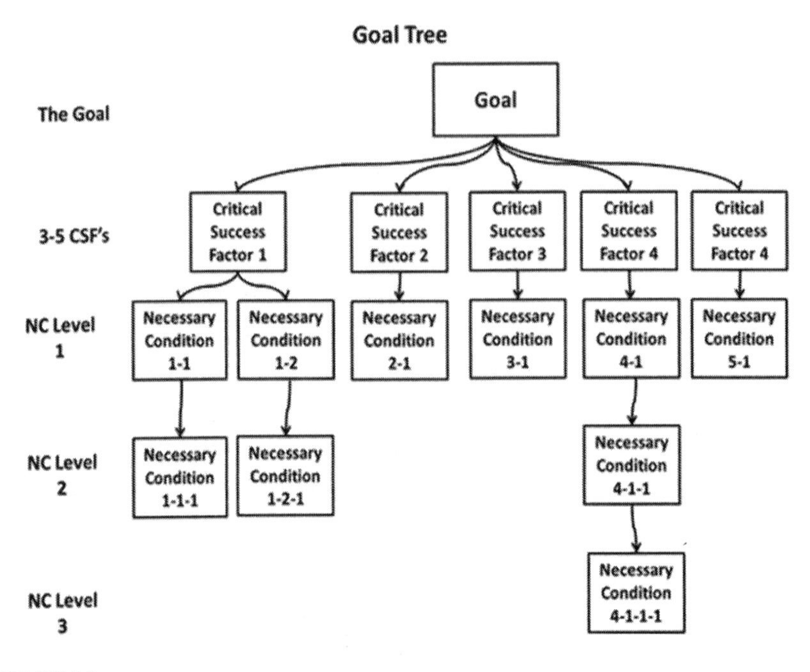

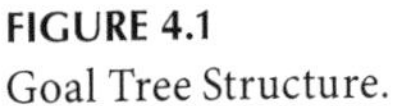

FIGURE 4.1
Goal Tree Structure.

The Goal and CSFs are worded as terminal outcomes, as though they were already in place, while the NCs are stated more as activities that must be completed.

Figure 4.1 is a graphic representation of the structure of the Goal Tree, with each structural level identified accordingly. The Goal, which is defined by the owners of the organization, sits at the top of the Goal Tree, with three to five Critical Success Factors directly beneath it. The CSFs are those critical entities that must be in place if the Goal is to be achieved. For example, if your Goal was to create a fire, then the three CSFs that must be in place are (1) a combustible fuel source, (2) a spark to ignite the combustible fuel source and (3) air with a sufficient level of oxygen. If you were to remove any of these CSFs, there would not be a fire. So, let's look at each of these components in a bit more detail.

THE GOAL

Steven Covey [3] suggests that to identify our goal we should "Begin with the end in mind," or where we want to be when we've completed our improvement efforts, which is the ultimate purpose of the goal. A goal is an end to which a system's collective efforts are directed. It's actually a sort of destination that implies a journey from where we are to where we want to be. Dettmer also makes it very clear that the *system's owner* determines what the goal of the system should be. If your company is privately owned, maybe the owner is a single individual. If there's a board of directors, they have a chairman of the board who is ultimately responsible for establishing the goal. Regardless of whether the owner is a single person or a collective group, the system's owner(s) ultimately establishes the goal of the system.

CRITICAL SUCCESS FACTORS AND NECESSARY CONDITIONS

In the Goal Tree, there are certain high-level requirements that must be solidly in place, and if these requirements aren't achieved, then we simply will never realize our goal. These requirements are referred to as *Critical Success Factors* and *Necessary Conditions*. Dettmer recommends no more than three to five CSFs should be identified. Each of the CSFs have some number of NCs that are considered prerequisites to each of the CSFs being achieved. Dettmer recommends no more than two to three levels of NCs, but in my experience, we have seen as many as five levels working well. While the Goal and the CSFs are written primarily as terminal outcomes that are already in place, the NCs are worded more as detailed actions that must be completed to accomplish each of the CSFs and upper-level NCs.

The relationship among the Goal, CSFs and the supporting NCs in this cascading structure of requirements represents what must be happening if we are to reach our ultimate destination. For ease of understanding, when I am in the process of constructing my Goal Trees, the connecting arrows are facing downward to demonstrate the natural flow of ideas. But when our structure is completed, I reverse the direction of the arrows to reveal the flow of results. In keeping with the thought of learning a tool and making it your own, I have found this works well for training purposes even though this is the complete opposite of Dettmer's recommendations for construction of a Goal Tree.

As we proceed, it's important to understand that the real value of a Goal Tree is its capability to keep the analysis focused on what's really important to system success. Dettmer [1] tells us that a "Goal Tree will be unique to that system and the environment in which it operates." This is an extremely important concept because "one size does not fit all." Dettmer explains that even two manufacturing companies, producing the same kind of part, will probably have very dissimilar Goal Trees.

CONSTRUCTING A GOAL TREE/INTERMEDIATE OBJECTIVES MAP

A Goal Tree could very quickly and easily be constructed by a single person, but if the system it represents is larger than the span of control of the individual person, then using a group setting is always better. So with this in mind, the first step in constructing a Goal Tree/IO Map is to clearly *define the system* in which it operates and its associated boundaries. The second consideration is whether or not it falls within your span of control or your sphere of influence. Defining your span of control and sphere of influence lets you know the level of assistance you might need from others, if you are to successfully change and improve your current reality.

Once you have defined the boundaries of the system, your span of control and sphere of influence you are attempting to improve, your next step is to *define the goal of the system*. Remember, we said that the true owner(s) of the system is/are responsible for defining the goal. If the true owner or owners aren't available, it is possible to articulate it by way of a "straw man," but even then, you need to get concurrence on the goal from the owner(s) before beginning to construct your Goal Tree. Don't lose sight of the fact that the purpose of the Goal Tree is to identify the ultimate destination you are trying to reach.

Dettmer tells us that the Goal Tree's most important function, from a problem-solving perspective, is that it constitutes a *standard of system performance* that allows problem solvers to decide how far off course their system truly is. With this in mind, your goal statement must reflect the final outcome and not the activities to get you there. In other words, the goal is specified as an outcome of activities and not the activity itself.

Once the goal has been defined and fully agreed on, your next order of business is to develop three to five Critical Success Factors that must be firmly in place before your goal can be achieved. As we explained earlier, the CSFs are high-level milestones that result from specific, detailed actions. The important point to remember is that if you don't achieve every one of the CSFs, you will not accomplish your goal.

Finally, once your CSFs have been clearly defined, your next step is to develop your *Necessary Conditions*, which are the simple building blocks for your Goal Tree. The NCs are specific to the CSF they support, but because they are hierarchical in nature, there are typically multiple layers of them below each of the CSFs. As already stated, Dettmer recommends no more than three layers for the NCs, but on numerous occasions I have observed as many as five layers working quite well. With the three components in view, you are now ready to construct your Goal Tree. Let's demonstrate this through a case study where a company constructed their own Goal Tree.

THE GOAL TREE CASE STUDY EXAMPLE

The company in question here is one that manufactures a variety of different products for diverse industry segments. Some orders are build-to-order, while others would be considered orders for mass production parts. This company had plenty of orders to fill, but unfortunately, they were having trouble not only filling them, but filling them on time. As a result, this company's profitability was fluctuating between making money one month, to losing money the next month. Because of this, the board of directors decided to make a leadership change and hired a new CEO to effectively "right the ship."

The new CEO had a diverse manufacturing background, meaning that in his career he had split his time between job shop environments and high-volume manufacturing companies. When the new CEO arrived, he called a meeting of his direct reports to not only meet them but to assess their proficiencies and capabilities. He soon realized that most of the existing management team had been working for this company for many years, and their skills appeared to be limited. Before arriving, the new CEO had concluded that the best approach to turning this company's profitability around and stabilizing it would be to use the Theory of Constraints Thinking Processes. But after meeting his new team and evaluating their capabilities, and since time was of the essence, he decided instead to use the Goal Tree to assess his new company and lay out an improvement strategy.

THE FIRST MEETING

The CEO's first order of business was to provide a brief training session on how to construct a Goal Tree for his new staff. The first step was to define the boundaries of their system, which ranged from receipt of raw materials from suppliers to shipping of their products to their customers. Within these boundaries, the team concluded that they clearly had defined their span of control, because they had unilateral change authority. They also decided that they could influence their suppliers, and the same to some extent with their customers, so their sphere of influence was also defined.

In advance of this first meeting with his staff, the CEO had met with the board of directors to determine what the goal of this company actually was. After all, he concluded, it's the owner or owners' responsibility to define the goal of the system, which was "Maximum Profitability." After reporting on his meeting with the board of directors to his team, and the goal they had decided on, the CEO posted the goal on the top of a flip chart as in Figure 4.2.

The CEO knew that the board of directors wanted maximum profitability both now and in the future, so he added the future reference to the Goal box. But before moving on to the CSFs, the CEO decided that it would be helpful if he explained the basic principles of both the concept of the system constraint, and how to use Throughput Accounting (TA). His staff needed to understand why focusing on the constraint would result in maximum throughput, but equally important, his staff needed to understand how the three components of profitability, Throughput (T), Operating Expense (OE) and Investment/Inventory (I), worked together to maximize profitability. With this in mind, the CEO began training his new team on the Theory of Constraints,

"Before we discuss Throughput Accounting, we need to learn about something called the Theory of Constraints." The CEO continued, "Consider this simple piping system used to transport water (Figure 4.3). The system is gravity fed, whereby water flows into Section A, then flows

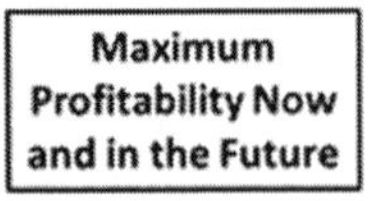

Maximum
Profitability Now
and in the Future

FIGURE 4.2
The Goal for Goal Tree.

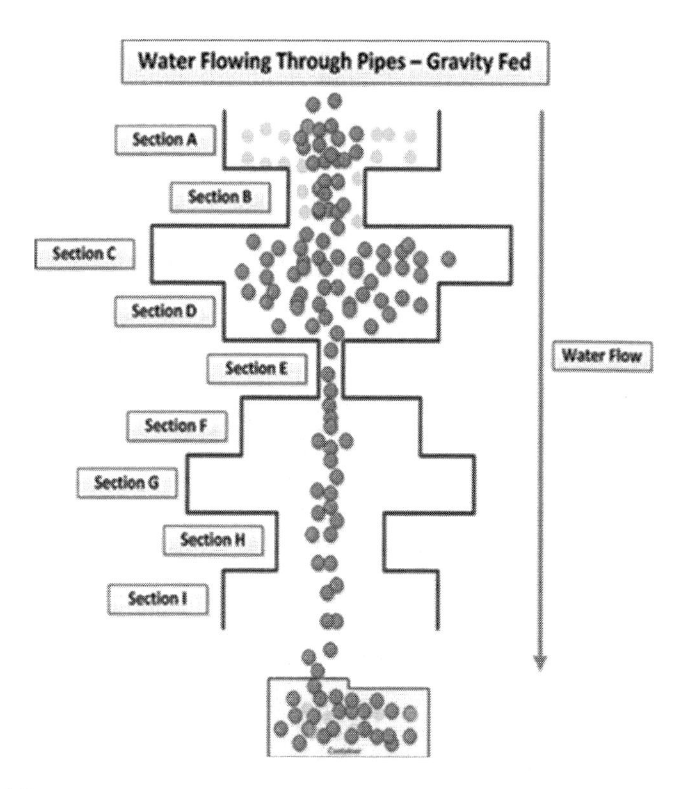

FIGURE 4.3
Basic Piping Diagram.

through Section B, then Section C and so forth until ultimately, the water collects in a receptacle immediately below Section I. It has been determined that the rate of water flow is insufficient to satisfy current demand and you have been called in to fix this problem." He then asked the group, "What would you do and why would you do it?"

After examining the diagram, the Operations Manager spoke up and said, "Because water is backing up in front of Section E, then Section E's diameter must be enlarged." The CEO smiled and asked, "Would enlarging the diameter of any other section of the piping system result in more flow of water through this system?" Everyone shook their heads, meaning that they all understood that only enlarging Section E's diameter would result in a higher flow rate of water.

The CEO then asked, "What factor determines how large Section E's diameter must be changed to?" The Quality Manager raised his hand and said, "That would depend upon how much more water was needed. In other words, what the demand was." Again, the CEO smiled, flashed a new

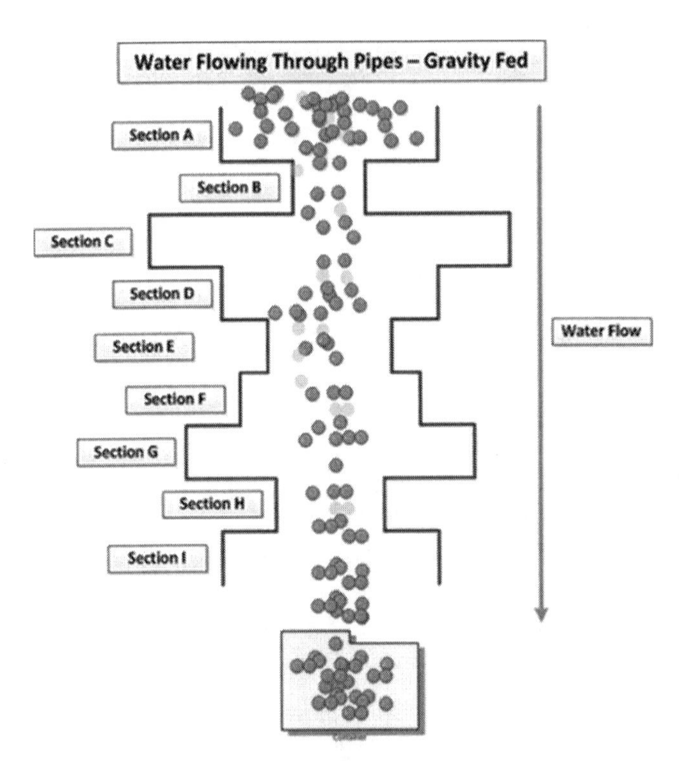

FIGURE 4.4
Piping Diagram with Constraint at New Location.

slide on the screen (Figure 4.4) and said, "So, let's see what happens when we enlarge the diameter of Section E of our piping system based upon the new demand requirement."

"As you can see, Section E's diameter has been changed, and water flow has increased. If you selected your new diameter based upon the new demand requirements, you will have 'fixed' this problem," he explained. "But what if there is another surge in water demand? What would you do?" he asked.

The Junior Accountant said, "You'd have to enlarge Section B's diameter and again its new diameter would be based upon the new demand requirement." The CEO then explained that Section E and now Section B are referred to as *system constraints* (aka bottlenecks). The inevitable conclusion in any business, is that the system constraint controls the flow and throughput within any system. He then asked, "So, how might this apply to our business?" and flashed another slide (Figure 4.5) on the screen.

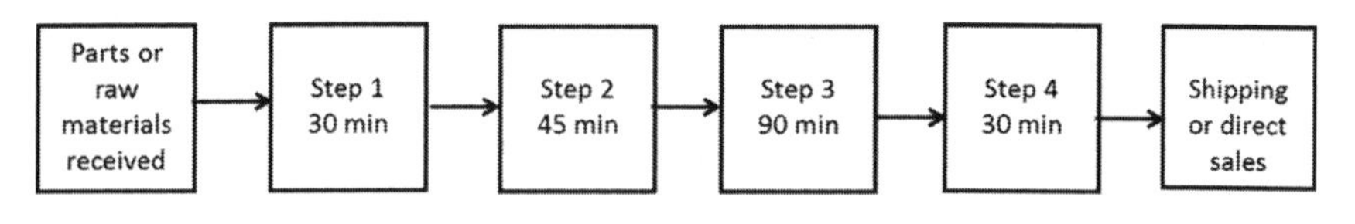

FIGURE 4.5
Process Example.

He continued, "For any type of business, there is a process that is at least similar to this figure. Materials, stock keeping units [SKUs] or parts are delivered to our business and enter into our manufacturing system. Step-by-step things happen to change the materials, until you have a finished product." The CEO continued,

> Parts or raw materials enter Step 1, are processed for 30 minutes and then passed on to Step 2. Step 2 processes what it has received from Step 1 for 45 minutes and then passes it on to Step 3. Steps 3 and 4 complete the processing and the finished product is either shipped to the company that ordered it, or directly sold to consumers. Suppose that you received orders for more parts. What would you do to increase the throughput of parts through this four-step process?

The CFO spoke up and said, "Because Step 3 takes the longest amount of time to complete, it is the system constraint, so the bottom line is that the only way to increase the throughput of this process is to reduce the time required in Step 3 (i.e. 90 minutes)." "You're right, and what would determine how much to reduce Step 3's processing time?" asked the CEO. The CFO replied, "Just like the piping diagram, it would depend on the demand requirements." The CEO than asked, "Would reducing the processing time of any other step result in increased output or sales?" Once again, the Junior Accountant raised her hand and said, "Absolutely not, because only the system constraint controls the output of any process!"

The CEO continued,

> Like us, many businesses are using manpower efficiency or equipment utilization to measure the performance of their processes and as a result of these performance metrics, they work to increase these two metrics. Since increasing efficiency is only achieved by running close to the maximum capacity of every step, guess what happens when this takes place? That is, Step 1 makes one part every 30 minutes and passes it on to Step 2 which takes 45 minutes to complete, etc.

Once again, the Junior Accountant raised her hand and said, "The process would be full of WIP [work-in-process] inventory." "Correct, and is there anything else that happens?" asked the CEO.

Without hesitation, the Junior Accountant said, "I would think that our cycle times would lengthen, and we'd probably be late on some orders." "Absolutely correct!" The CEO replied.

The CEO explained, "After the first eight hours, this is what this process looks like (Figure 4.6). WIP begins to accumulate—total processing time increases—on-time delivery deteriorates—customers get frustrated. Frustrated because of our company's inability to ship product on time, which negatively impacts our future order rate. And as time passes, after three eight-hour days, WIP levels continue to grow, negatively impacting flow, and unless something changes, the system becomes overwhelmed with WIP (Figure 4.7). This increase in WIP extends processing time even further which negatively impacts on-time delivery and customers end up threatening to take their business elsewhere. So, what is the answer?"

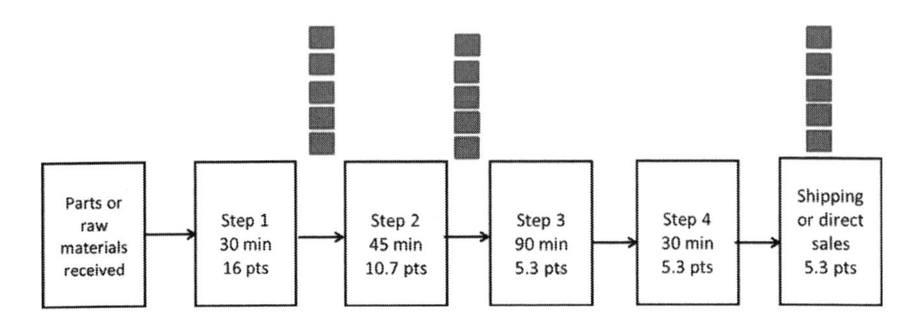

FIGURE 4.6
Process with WIP Inventory after One Day.

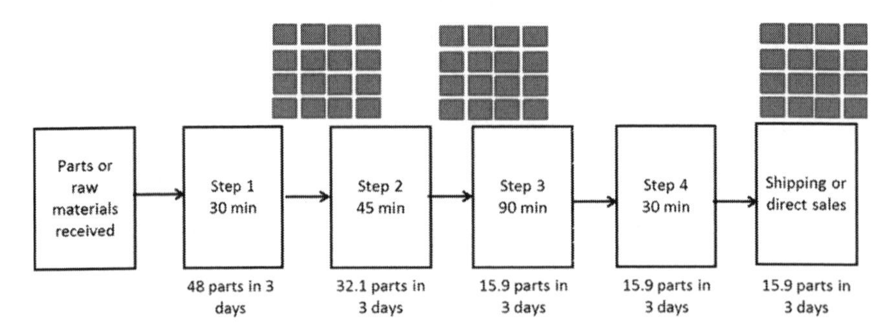

FIGURE 4.7
Process with WIP Inventory after Three Days.

The CEO continued, "As you have just witnessed, the performance metrics efficiency or equipment utilization, both have negative consequences, so maybe they're not such good metrics for us to use after all, at least not in non-constraints? They are both excellent metrics to use to drive our constraint's output, but not in our non-constraints. So, in order to avoid an explosion of WIP, doesn't it make sense that Steps 1 and 2 should be running at the same speed as the constraint (i.e. one part every 90 minutes)?" "In order to increase the output rate of this process, Step 3's processing time must be reduced because it is the system constraint. What happens if we focus our improvement efforts on Step 3 only?" The Operations Manager spoke up and said, "By focusing our improvement efforts on Step 3 only, we get an immediate increase in the throughput of this process." The CEO smiled and said, "This is the essence of the Theory of Constraints … it provides the necessary focus for improvement efforts. So now, let's move on to a comparison of Cost Accounting and Throughput Accounting."

The CEO explained that Throughput (T) was equal to Revenue (R) minus Totally Variable Costs (TVC) and that Net Profit was equal to Throughput minus Operating Expense (OE), or T – OE. Finally, he explained that Return on Investment (ROI) was equal to NP/I, where I is the Inventory value. With this brief training behind them, he then challenged his staff to tell him what they needed to have in place to satisfy this profitability goal, both today and tomorrow. That is, what must be in place to maximize Net Profit, now and in the future?

After much discussion, his staff offered three Critical Success Factors which the CEO inserted beneath the Goal in the Goal Tree (Figure 4.8). After learning the basics of TOC's concept of the constraint, and basic Throughput Accounting, his staff knew that because they needed to increase Net Profit (T – OE), then maximizing Throughput had to be one of the CSFs. They also concluded that, to maximize Net Profit, minimizing OE had to be another CSF. And finally, because ROI was equal to Net Profit divided by their Investment (i.e. NP = (T ÷ I), they needed to include minimum investment (i.e. inventory) as one of the CSFs. The CEO felt very good about the progress they had made with their first Goal Tree, but it was time for lunch, so they decided to break and come back later to complete the Goal Tree.

When his staff returned from lunch, they reviewed what the CEO had presented on TA, just so it was fresh in their minds, as they began again to review and construct their Goal Tree. The CEO started, "In order to

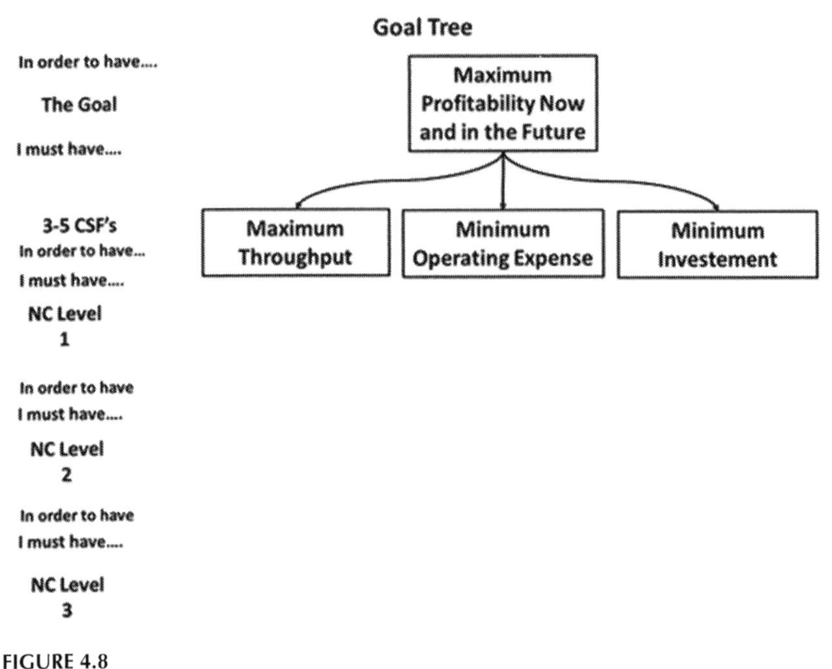

FIGURE 4.8
Goal Tree with Goal and CSFs.

maximize profitability now and in the future, we must have maximum Throughput, minimum operating expense and minimum investment, which is mostly inventory." "Are there any others?" he asked. His staff looked at each other and agreed that these are the three main CSFs.

The CEO knew that what was needed next were the corresponding Necessary Conditions, so he started with Maximum Throughput. "In order to have maximum Throughput, what do we need?" His CFO put his hand up and said, "We need to maximize our revenue stream." Everyone agreed, but the Junior Accountant immediately raised her hand and said, "That's only half of it!" The CEO and CFO looked at her and said, "Tell us more." She explained, "Well, you explained that Throughput was Revenue minus Totally Variable Costs, so minimal Totally Variable Costs has to be a Necessary Condition too." The CEO smiled and said, "So, let me read what we have so far. In order to have maximum Throughput, we must have maximum Revenue and minimal TVCs," and everyone agreed.

The CEO continued, "In order to maximize Revenue, what must we do?" The Operations Manager said, "We must have satisfied customers," and before he could say another word, the Marketing Director added, "We must also have sufficient market demand." The CEO smiled, scanned the room for acceptance again, and added these two NCs to the Goal Tree (Figure 4.9). The CEO thought to himself, I am so happy that I chose to use the Goal Tree rather than the full Thinking Process analysis.

The CEO then said, "Let's stay with the satisfied customers NC. In order to have satisfied customers, we must have what?" The Quality Director raised his hand and said, "We must have the highest quality product." The Logistics Manager added, "We must also have high on-time delivery rates." And before the CEO could add them to the tree, the Customer Service Manager added, "We must also have a high level of customer service." The CEO smiled again and said, "Slow down, so I don't miss any of these, everyone." Everyone laughed. The CEO looked at the lower level NCs for satisfied customers and asked if they needed anything else. Everyone agreed that if they had the highest quality product, with high on-time delivery rates and a high level of customer service, then the customers should be highly satisfied.

The CEO continued on beneath the CSF for Maximum Throughput and asked, "So, what do we need to supplement or support sufficient market demand?" The CFO said, "We need a competitive price point and by the

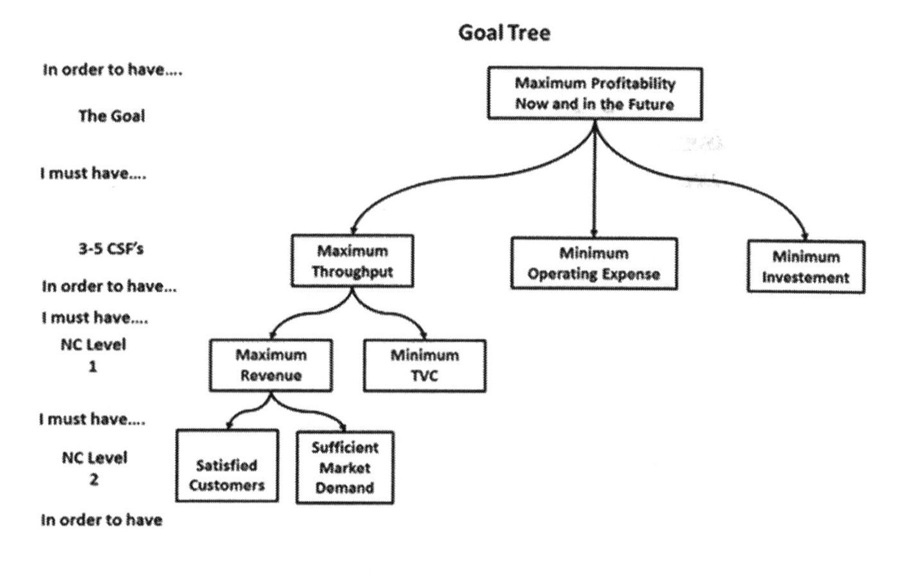

FIGURE 4.9
Goal Tree with Goal, CSFs and Some NCs.

way, I think that would also help satisfy our customers." The CEO added both NCs and connected both of them to the upper-level NC of sufficient market demand. The CEO stepped back and admired the work they had done so far, but before he could say anything, the Sales Manager said, "If we're going to have sufficient market demand, don't you think we also need effective sales and marketing?" Again, everyone nodded their heads in agreement, so the CEO added that NC as well.

Before the CEO could say anything more, the Junior Accountant raised her hand and added, "I was thinking that three of the ways we could have effective sales and marketing would be related to the three lower level NCs assigned to satisfied customers. I mean, can we do that in a Goal Tree?" The CFO was the first person to speak and he added, "I think that's a fantastic idea!" The CEO thanked her and added the connecting arrows.

The CEO then said, "Great job so far, but what's a good way for us to minimize TVC?" Without hesitation, the Quality Manager said, "That's easy, we need to minimize our scrap and rework." The Quality Manager then said, "I think that would also be an NC for one of our other CSFs, minimum operating expense." Everyone agreed, so the CEO added both the NC and the second connecting arrow. Once again, the Junior Accountant raised her hand and added, "I think that we should add another NC to the CSF, minimum operating expense, and that we should say something like optimum manpower levels and maybe also minimized overtime." The CEO smiled and added both of the NCs to the tree.

"So, what about our CSF, Minimum Investment?" asked the CEO. The Plant Manager raised his hand and said, "How about minimized WIP and Finished Goods inventory?" The CEO looked for objections, but when nobody objected, he added it to the tree. He then asked, "What about an NC underneath that one?" The Plant Manager looked at him and said, "We need to synchronize our production around the constraint and demand." "What do you mean?" asked the CEO. "I mean we need to stop producing parts on speculation, and start building based on actual orders. I've been reading about TOC version of scheduling referred to as Drum Buffer Rope and I think we need to move in that direction," he added. And with that, the CEO added his comments to the Goal Tree/IO Map.

When he was finished adding the new items to the Goal Tree (Figure 4.10), he turned to the group and began clapping his hands in appreciation for their effort. He explained, "I've been doing this for quite some time now, but I have never seen a group come together more than you have today." He added, "I was a bit apprehensive when we began

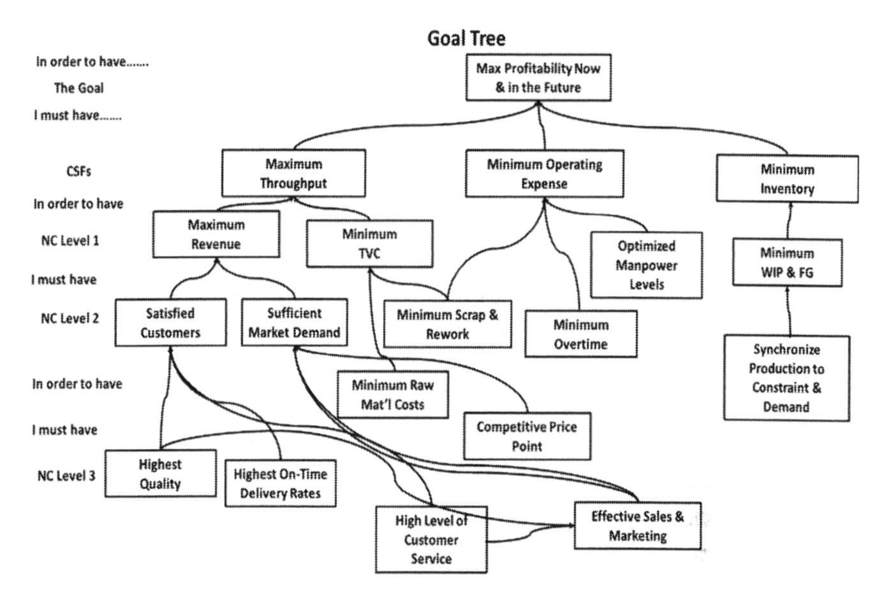

FIGURE 4.10
Completed Goal Tree.

today, that maybe some of you would push back and not contribute, but I was totally wrong."

The CFO raised his hand and said, "For me, I have never seen this tool before, but going forward, I will be using it a lot. In fact, I was thinking that this tool can be used to develop individual department improvement plans." "Tell us more," said the CEO. "Well, if an NC, for example, applies mostly to a specific group like Production or Sales and Marketing, then that could be seen as the Goal for that group. I'm very happy to have been here today to complete this exercise," said the CFO. Everyone else agreed with him. The CEO then said, "Ladies and gentlemen, this exercise is not over yet." "What else is there to do?" came a question from the CFO. "We'll get back together tomorrow morning and I'll explain the next steps," he explained.

USING THE GOAL TREE AS AN ASSESSMENT TOOL

Bright and early the next morning, the executive team began filing into their conference room, full of anticipation on just what they would do with their completed Goal Tree. The CEO hadn't given them any instructions on how to prepare for today's work, so they were all eager to have the events of the day unfold. When everyone was seated, the CEO welcomed them and offered his congratulations again on the great job they had done the day before. "Good morning everyone," he said as everyone responded with a "good morning" back to him. As he scanned the room, he noticed that there was one person missing, the Junior Accountant. When he asked the CFO where she was, he explained that she was working on the monthly report and wouldn't be joining them today. The CEO looked the CFO square in his eyes and told him that nothing was more important than what they were going to do today. "Go get her!" he stated emphatically. The CFO left and returned minutes later with the Junior Accountant, and the CEO welcomed her. He then said, "We created this as a complete team and we're going to finish it as a complete team."

The CEO explained, "When the Goal Tree was originally created by Bill Dettmer [1], it was to be used as a precursor to the creation of a Current Reality Tree (CRT). That is, he used it as the first logic tree in TOC's Thinking Processes to help create the CRT." He continued, "And although I fully support this approach, I have found a way to use the Goal Tree to accelerate the development of an improvement plan." The CEO passed out copies of the completed Goal Tree and began.

"I want everyone to study our logic tree, focusing on the lower level NCs first," he explained. "As we look at these NCs, I want everyone to think about how we are doing with each of these," he continued. "By that I mean, is what we said is needed to satisfy a CSF or upper-level NC in place and functioning as it should be." "We're going to use a color-code scheme to actually evaluate where we stand on each one," he said. "If you believe that what we have in place is good and that it doesn't need to be improved, I want you to color it green. Likewise, if we have something in place but it needs to be improved, color it yellow. And finally, if an NC is either not in place or is not 'working' in its current configuration, color it red," he explained. "Does everyone understand?" he asked, and everyone nodded in agreement. "It's important that we do this honestly, so be truthful or this exercise will all be for nothing."

The CFO raised his hand and asked, "How will we use our color-coded tree?" "Good question," said the CEO. "Once we have reviewed our Goal, CSFs and NCs, we will start with the red entities first and develop plans to turn them into either yellows or greens. Likewise, we'll then look at the yellows and develop plans to turn them into green ones," he explained. As he was explaining his method the CEO could see heads nodding in the affirmative, meaning that everyone understood his instructions. With that, the CEO passed out green, yellow and red pencils. "I want everyone to do this individually first, and then we'll discuss each one openly, until we arrive at a consensus," he explained. "While you're considering the state of each entity, I also want everyone to also think about a way we can measure the status of many of these in the future," he said. "I'll be back in a couple of hours, so please feel free to discuss your color selections as a group," he added. With the instructions complete, the team began reviewing their Goal Tree and applying the appropriate colors to each entity.

Right on schedule, the CEO returned and asked how the session was coming. The Plant Manager spoke first, "I was amazed at how much disagreement we had initially, but after we discussed each item, we eventually came to an agreement on how we believe we're doing." The CFO jumped into the conversation and added, "I was amazed at how we came together as a team, just by creating our Goal Tree." "I have to admit that when you told me to go get our Junior Accountant, I was a bit taken back. But at the end of the day, she was a very important addition to this team," he added. And with that, the Junior Accountant was somewhat embarrassed, but thanked the CFO for recognizing her contribution to the effort.

"So, where is it?" asked the CEO. "Where is your finished product, your Goal Tree?" The CFO went to the flip chart and there it was (Figure 4.11). The CEO then asked, "Did you also discuss what kind of metrics we might

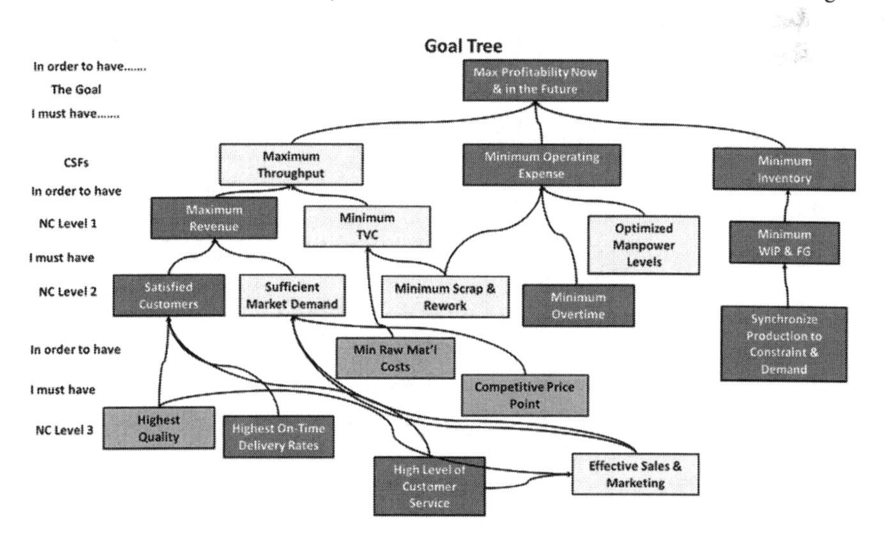

FIGURE 4.11
Completed Goal Tree after Analysis. Red is represented by dark gray, yellow by medium gray and green by light gray.

use to measure how we're doing?" "Yes, we did," said the CFO. "And?" the CEO asked. "We need to do more work on that," he answered. "So, what's next?" asked the CFO. After studying the finished product, the CEO thanked everyone for their effort and then said, "Let's take a break and come back later, and I'll explain how we can use this tree to develop our final improvement plan."

The team reassembled later that day to discuss their next steps. Everyone seemed enthusiastic about what they would be doing going forward. When everyone was seated, the CEO turned to the group and asked, "So, how does everyone feel about this process so far?" The Plant Manager was the first to respond, "I can't speak for anyone else, but the development of the Goal Tree was a real eye-opener for me. I never imagined that we could have analyzed our organization so thoroughly in such a short amount of time. I mean think about it, when you add up the total amount of time we've spent so far, it's not even been a full day's work!" As he spoke, everyone was nodding their heads in agreement.

The CFO was next to speak and said, "I can absolutely see the benefit from using this tool, and one of the things that impressed me the most is that everyone contributed. But what really captivated me is that for the first time since I started working here, we actually are looking at the system rather than isolated parts of it. One of the things that I will take away from this is that the total sum of the localized improvements does not necessarily result in an improvement to the system. The Goal Tree forces us to look at and analyze all of the components of our organization as one entity."

"OK, let's get started," said the CEO. "Today we're going to plan on how turn our problem areas, those we defined in red, into—hopefully—strengths. Does anyone have any ideas on how we can turn our bottom three reds into either yellows or greens?" "In other words, what can we do that might positively impact delivery rates, customer service and synchronize production to the constraint and demand?" he asked.

The Plant Manager was the first to speak. "If we can come up with a way to schedule our production, based upon the needs of the constraint, it seems to me that we could really have a positive result for on-time delivery rates and at the same time it would reduce our WIP and FG levels?" he said, more in the form of a question. The CFO then said directly to the Plant Manager, "Since you mentioned Drum Buffer Rope yesterday, I've been reading more about it and it seems that this scheduling method is supposed to do exactly what you just described."

The CEO responded by saying, "He's right, DBR limits the rate of new product starts, because nothing enters the process until something exits the constraint." "So, let's look at what happens to the reds and yellows, if we were to implement DBR," he added and pointed at the Goal Tree up on the screen. "The way I see it is, if we implement DBR, we will minimize WIP. If we minimize WIP, we automatically minimize FGs, which minimizes our investment dollars, which positively impacts our profitability," he explained enthusiastically. "We should also see that on-time delivery rates jump up, which should result in much higher levels of customer satisfaction," he added. "This should also allow us to be more competitive in our pricing and stimulate more demand and with our ability to increase throughput, we will positive impact profitability," he explained. The Junior Accountant then said, "Last night I read more about the Theory of Constraints and it seems to me that one thing we could do is stop tracking efficiency in our non-constraints and if we do that we should also reduce our WIP."

The Quality Director spoke up and said, "I'm thinking that if we effectively slow down in our non-constraints, we should see our scrap and rework levels improve significantly, because our operators will have more time to make their products. And I also believe that we should implement TLS." "What is TLS?" asked the CFO. "It's an improvement method which combines the Theory of Constraints, Lean and Six Sigma," the Quality Director explained. "This improvement will reduce our scrap and rework levels and, in conjunction with DBR, will reduce both our operating expenses and TVC. The combination of these improvements will both contribute to our profitability," he added.

"One other thing is that we should see our overtime levels drop, which will also improve profitability," said the CFO. "I am just amazed that by making these three basic changes, we could see a dramatic financial improvement," he added. The team continued working on their Goal Tree until it was complete (Figure 4.12).

The stage was set for major financial gains by first developing their cause and effect relationships, and by looking at their organization as a system rather than making improvements to parts of it. That's a very important message for everyone to glean from all of this. Not all improvement efforts will happen rapidly like they did in this case study, but it is possible to make rapid and significant improvements to your organization by looking at it from a holistic point of view. The fact is, isolated and localized improvements will not typically result in improvement to the system. So, let's get back to our case study.

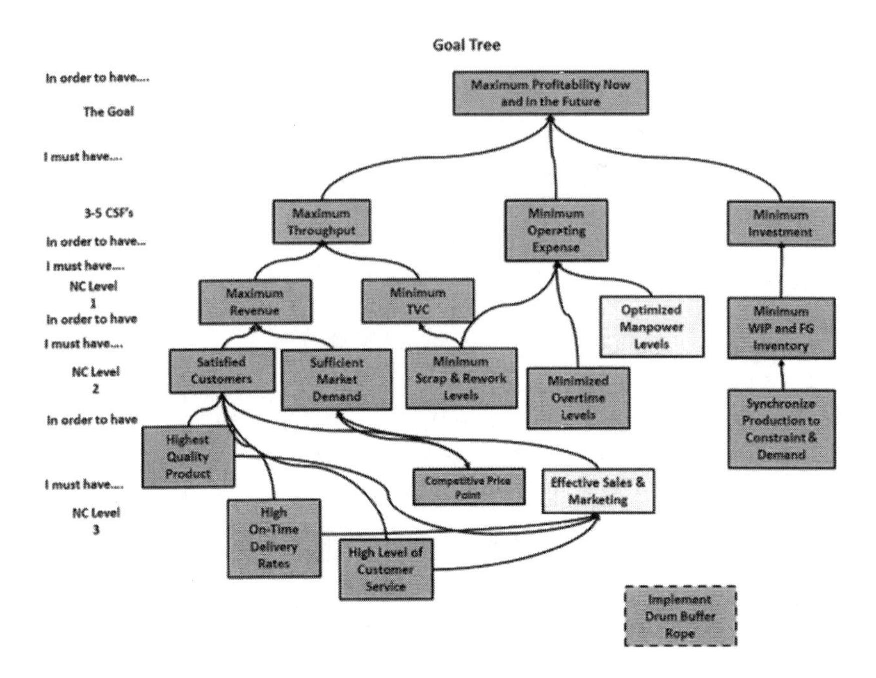

FIGURE 4.12
Completed Goal Tree after First Improvement. Red is represented by dark gray, yellow by medium gray and green by light gray.

PERFORMANCE METRICS

The CEO began to explain, "Before we develop our performance metrics, let's first discuss the purposes of an organization's performance metrics. In general, we need some type of feedback mechanism that tells us how we're doing. A way to be able to know that the direction we're traveling is on course. That is, in the event that we need to make any mid-course corrections. These performance metrics should be system related, in that they tell us how the system is performing rather than how individual processes are functioning. Remember, our focus is on system performance, and not individual performance. So, think about what our performance metrics should be?" "But before we answer that question, let's talk about their purpose," he added.

The CEO continued, "Performance measures are intended to serve at least six important functions or roles," as he wrote them on the flip chart:

1. First, and foremost, the measures should stimulate the right behaviors.

2. The performance measures should reinforce and support the overall goals and objectives of the company.
3. The measures should be able to assess, evaluate and provide feedback as to the status of the people, departments, products and the total company.
4. The performance measure must be translatable to everyone within the organization. That is, each operator, manager, engineer, etc. must understand how their actions impact the metric. Performance metrics are intended to inform everyone, not just the managers!
5. The performance metrics chosen should also lend themselves to trend and statistical analysis and, as such, they shouldn't be "yes or no" in terms of compliance.
6. The metric should also be challenging, but at the same time be attainable. There should be a stretch involved. If it's too easy to reach the target, then you probably won't gain nearly as much in the way of profitability. If it's too difficult, then people will be frustrated and disenchanted.

The CEO continued, "So with these functions in mind, let's now look at how we can use our Goal Tree to create our series of performance metrics." "If we use the Goal Tree/IO Map as our guide, we should start with our goal, *Maximize Profitability Now and In the Future*, and create our first tracking metric," he explained. "Earlier in our discussion, I introduced you to Throughput Accounting, which defined Net Profit as Throughput minus Operating Expense or NP = T − OE. The metric of choice for this Goal Tree then, should be NP which we insert into our 'goal box.' In addition, I prefer to give most of the metrics a target to shoot for. With this metric, I believe that our Net Profit should be greater than 25 percent (NP >25%)," he stated. "We must then look at each CSF and NC and select appropriate performance metrics and targets for as many as might be appropriate. Keep in mind that not every box will have a defined metric, but let's get as many as we can," the CEO explained. The CEO then told the executive team that he wanted them to work on the rest of the metrics as a team and that he would return later.

Because the operational status of companies varies from company to company, there is no standard set of metrics and targets to recommend. But for the company in this case study, the team stayed focused and was able to identify appropriate metrics and targets. They started with the

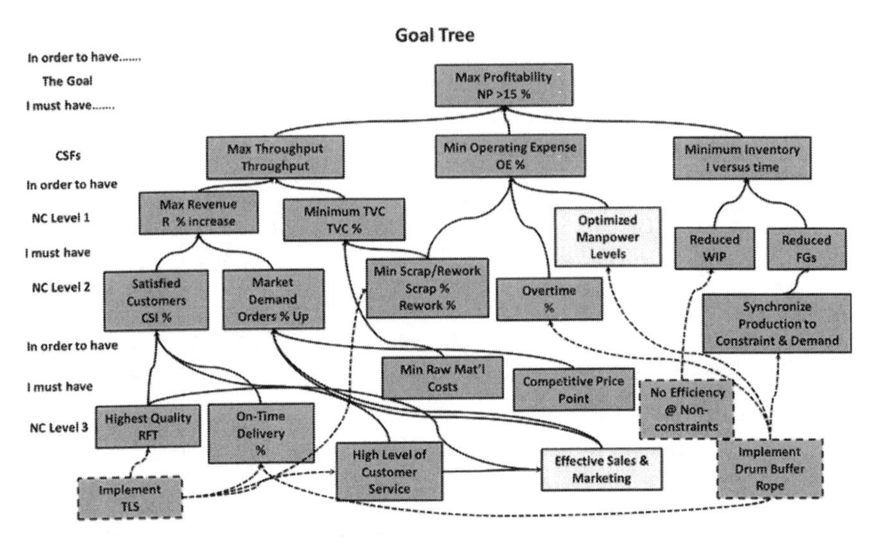

FIGURE 4.13
Completed Goal Tree after Several Improvements.

Goal, then worked through the CSFs, and then onto the NCs. Let's get back to the case study and see what they were able to do.

The CFO was the first to speak and said, "It's clear to me that having performance metrics for the three CSFs is imperative, since they are the three components of profitability." The team decided to first determine which entities could actually have metrics tied to them. After they had determined all of the metrics, they would then develop targets for each specific performance metric. The next figure (Figure 4.13) is the Goal Tree with appropriate metrics defined by this executive team.

All of the team members contributed to this effort, and all were amazed at the finished product they had developed. As they were admiring their Goal Tree with the performance metrics they intended to track, the CEO entered the room. He studied the completed Goal Tree and then moved to the front of the room. He thanked everyone for their effort and told them that they could return to their offices. The CFO stood and said, "With all due respect sir, we haven't finished yet. We have identified the key metrics we want to track, but we must now develop our targets for each metric." The CEO just smiled and told them he would be back later to see their finished product.

The CFO stood, faced the group and asked, "Where do you think we should start? I mean should we start at the top with the Goal or should we start at the bottom and work our way up?" The Junior Accountant raised her hand and said, "I think we should start at the lower levels and work our way to the top." "Why do you feel that way?" asked the CFO. "If we follow the direction of the arrows and then set, and reach, our target, then the level directly above will be the net result of our efforts," she replied. "Could you give us an example?" asked the CFO. "OK, for example, if we set our target for Highest Quality, Right the First Time (RFT) at greater than 99 percent and we achieve it, plus, our on-time delivery rate to 99 percent and we achieve that, then we have a great chance of having our Customer Satisfaction Index (CSI) be greater than 99 percent. So, it's kind of like sufficiency type logic using if-then statements," she explained.

The Quality Director then spoke up and said, "I can see your point, but I can also make the argument based on necessity-based logic. By that, I mean we could start with the Goal and give it a target of 25 percent. So, with necessity-based logic, we could say that in order to have a profitability of 25 percent, we must have a Throughput improvement of at least 20 percent, while holding our operating expenses to less than 10 percent and holding our on-hand inventory to less than one day." The CFO re-entered the conversation and said, "I see both points of view, but I must tell you I like following the direction of the arrows on our Goal Tree. I say this because, when we implement Drum Buffer Rope and TLS, we drive improvement upward and our metrics respond to what we're doing." The Junior Accountant then said, "I don't think it matters which direction we go, because at the end of the day, the metrics will tell us how we're doing." Everyone nodded their heads in agreement, and they got busy setting their targets for each defined performance metric. When they finished, they recorded them in a PowerPoint slide for presentation to the CEO.

A short while later, the CEO returned to the conference room to find the completed Goal Tree, with metrics and targets posted on the screen at the front of the conference room (Figure 4.14). He studied it, turned to the group and asked someone to explain it to him. The CFO turned to the Junior Accountant and said, "I think since you contributed most to our success, that you should be the one to do that." The Junior Accountant just smiled and said she would be happy to, and did so with confidence, agility and a seemingly true understanding of this new tool. She finished her presentation by telling the executive team that this tool will serve her well in her new position as CFO with her new company and everyone gasped in disbelief.

The Goal Tree is not only an amazingly simple tool to learn, but in my experiences, it's a tool that most people feel comfortable using. As you

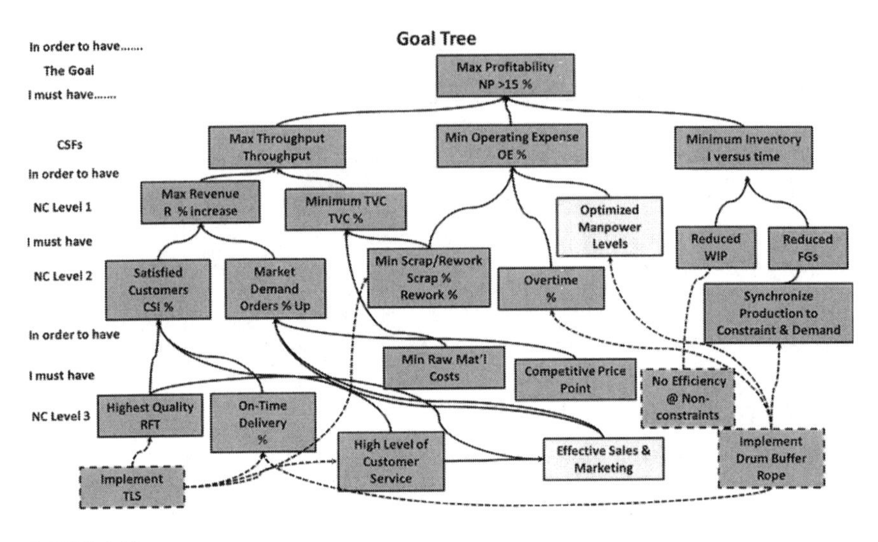

FIGURE 4.14
Final Goal Tree with All Improvements in Place.

learned, in a very short amount of time, this team not only learned how to create their Goal Tree, they were able to use it to develop their strategic and tactical improvement plan. I am forever grateful to Bill Dettmer for providing us with this amazing tool, and I encourage everyone to read Bill's book, *The Logical Thinking Process—A Systems Approach to Complex Problem Solving.*

In the next chapter, we will present the Logical Thinking Processes, so that you will have a good understanding of just how involved this set of tools really is. This next chapter will also demonstrate why we have elected to present the Goal Tree first.

REFERENCE

1. Dettmer, H. William, *The Logical Thinking Process—A Systems Approach to Complex Problem Solving*, American Society for Quality Control Press, Milwaukee, WI, 2007.

5

The Logical Thinking Process

INTRODUCTION

As mentioned in the last chapter, we're going to focus on another side of the Theory of Constraints (TOC), the Logical Thinking Processes (LTPs). The LTPs are made up of six logic trees and the "rules of logic" that direct their construction. These six logic trees are as follows:

1. The Intermediate Objectives Map/Goal Tree
2. The Current Reality Tree
3. The Conflict Resolution Diagram
4. The Future Reality Tree
5. The Prerequisite Tree
6. The Transition Tree

Before we get into how to construct each of these logic trees, let's talk about the purpose of each type of logic tree and the whole idea of using logic tools in general.

If you're a manager, do you have a good understanding of your company's goal? Intuitively you know that your company's goal is to make money, especially for the stockholders. But being a manager, you know that it's more than just making money, because there are other important things to consider. For example, things like having a competitive edge, having enough market share, having high levels of customer satisfaction, and first-time quality levels, and what about costs? Aren't they important as well? Not all of these items can be classified as goals, but we know that without them we wouldn't be making money in the long run. So, it seems that we need an orderly way to classify different things that are critical to our long-term success, and we need a roadmap on how to get to where we want to go. So where do we start? If you're a fan of Stephen Covey, he would tell you to "begin with the end in mind." For us, Steven Covey got it absolutely right, beginning with the end in mind, with the end being achievement of the goal.

Achievement of the goal of an organization must be considered as a journey, simply because there are intermediate steps along the way that must be achieved first. In TOC, we refer to these as Critical Success Factors (CSFs). But even before we achieve these CSFs, there are necessary conditions (NCs) that must be met first. As we discussed in the last chapter, the Goal, CSFs and NCs arrange themselves as a nested hierarchy. The Goal Tree/Intermediate Objectives Map (IO Map) is the tool we use to determine these three entities. Dettmer [1] refers to the IO Map as a "destination finder," and rightly so. As explained in the last chapter, and to review, the IO Map/Goal Tree begins with a clear and unmistakable statement about the purpose of the organization, the *Goal*. Next on the hierarchy are several CSFs followed by NCs. These three elements are structured as such and represent what should be happening in your organization.

Dettmer explains that an IO Map is really intended to create a firm baseline or standard of what should be happening, if the system is going to successfully reach its goal. To determine how you're actually doing, you must have a good understanding of what you should be doing.

THE LOGICAL THINKING TOOLS

In our last chapter, we introduced you to the Goal Tree/IO Map, as the first of the TOC Thinking Process Tools. Now let's continue with the second of these tools, referred to as the Current Reality Tree (CRT). The CRT is a sufficiency-based logic tree (i.e. uses if-then statements) and is extremely useful when trying to solve organizational problems.

The Theory of Constraints is systemic in nature and strives to identify those few constraints that limit the organization's success, in terms of moving in the direction of its goal. It's important to keep in mind that most organizations function as systems, rather than as just processes. Goldratt introduced his 5 Focusing Steps, plus what he calls a logical thinking process. Goldratt then taught us that good managers must answer five important questions to be successful (Figure 5.1):

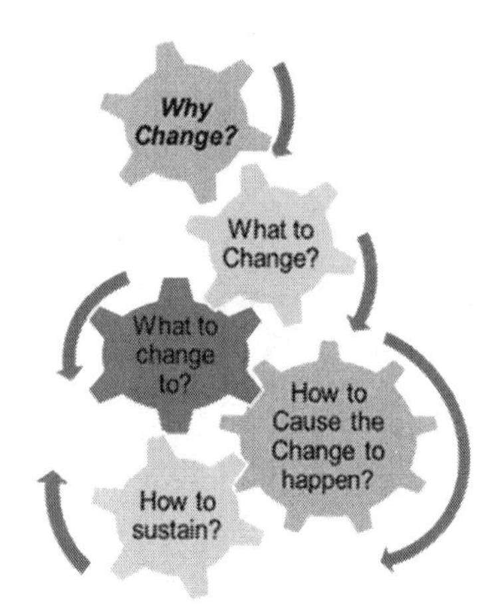

FIGURE 5.1
Five Key Questions.

1. Why change?
2. What to change?
3. What to change to?
4. How to cause the change to happen?
5. How to sustain?

In my experiences, answering these five questions is the hallmark of a successful system's transformation. What's the answer to the first question, *Why should we change?* If you're happy with your current performance, then is there a reason to change? The answer is a resounding yes! You should never be satisfied with your current results, because today's performance will simply not be good enough for tomorrow. You may not think you need to change, but in today's competitive world, change is a necessity.

The second question is *What do you change?* It should be apparent by now that changing everything will not result in significant bottom-line improvement. When you have limited resources, like many of your companies do, you must *focus* your collective efforts on the right thing to change and then *leverage* it. Unless the question of "what to change" is explicitly addressed at the strategic, tactical and operational levels, any significant and sustainable benefit of the "improvement" is usually a matter of chance.

What do you change to? Even if you correctly decide what to change, how do you know what to change it to? Without development of a clear, cause and effect understanding of the problems you face, there is no way to predict the results with any confidence. Hopefully, by now you have a better understanding of your system's leverage point.

How do you cause the change to happen? Even if you successfully determine what to change and what to change to, what do you have to do to successfully make the change happen? Putting in place a significant and sustainable improvement effort requires careful planning, support of the necessary players and tracking of expected outcomes along the way to effectively carry out those plans.

The final question that must be answered is, *How do you sustain the gains?* It's great to make all these improvements, but what must you do to guarantee that the improvements you have made will continue to remain in effect?

Let's begin to answer these five questions in more detail. Correctly answering these five questions will take you to the improvement promised land! If you don't learn anything else from this book, remember that the key to improvement is locating your *system's leverage point*, and focusing improvements there, until a new leverage point appears. And when it does, you simply move your improvement efforts to a new location.

Before we begin our LTP discussion, we want to interject with what we refer to as the "7 Must Haves" for improvement. We must have the following:

1. A *focusing mechanism* to assure that you are improving the correct part of your organization, its *leverage point.*
2. A methodology that first identifies value-added, non-value-added and non-value-added but necessary activities within your organization and then demonstrates ways to remove or reduce their effects.
3. A methodology that identifies *excessive* and *uncontrolled variation* within your organization and then demonstrates ways to reduce and control it. It is important to remember that not all variation can be *removed*, but it can be reduced and then *controlled.*
4. A way to *link* all three of these methods into a *single* methodology.
5. The old "*command and control*" management style must be removed and replaced with the *new role of management*, the removal of any and all obstacles that stand in the way of continuous improvement.
6. The true, *active involvement* of your company's true subject matter experts (your front-line employees) and the willingness to both

listen to their ideas and then *implement* them, as long as safety, company policies or customer contractual obligations aren't compromised.

7. The mindset of being a *satisficer* rather than an *optimizer.*

What the heck is a satisficer and an optimizer? Nobel laureate H. A. Simon recognized many years ago (circa 1957) a management situation that, according to Simon, caused significant decision-making hardship. Simon claimed that decision makers, like executives, managers, engineers, etc., were trying to become *optimizers* while making decisions. According to Simon, optimizers are decision makers who are always in search of the *best possible solution,* without considering time or resource constraints. To achieve the best possible decision, the optimizer gathers all of the information needed to build a model that will allow them to choose the best alternative. The problem with being an optimizer is that this approach requires significant amounts of time, effort and money to achieve critical solutions, because you're always looking for the perfect solution.

Satisficers are decision makers who are satisfied with a reasonable solution, that will result in significant improvement to the system, rather than waiting for the perfect solution. According to Simon, the satisficer sets a *level of aspiration*, a threshold or objective to be achieved. The satisficer's objective is not to maximize or minimize some performance measure, but to achieve a solution that will improve the measure at or beyond their predefined level that they had set. When this level has been achieved, the satisficer sets a new target to surpass and the process repeats itself. This sort of stair-step approach to improvement, keeps the organization moving in a positive direction, without waiting to achieve perfection. So, which is the best way in terms of continuous improvement?

Herb Simon suggested that decision makers should behave as satisficers and that they should seek to reach a satisfactory solution and not an optimal one. A satisficer wins by complying with two basic principles. First, set a high enough level of aspiration, consistent with market conditions, competition and investor expectations, and second, adopt an approach of continuous improvement. The point is, if your management style is one that searches for an *optimal solution*, valuable time is lost making improvements. Being a satisficer, and using a stair-step approach, always works better (Figure 5.2)! Now. Let's get back to our explanation of the Logical Thinking Processes.

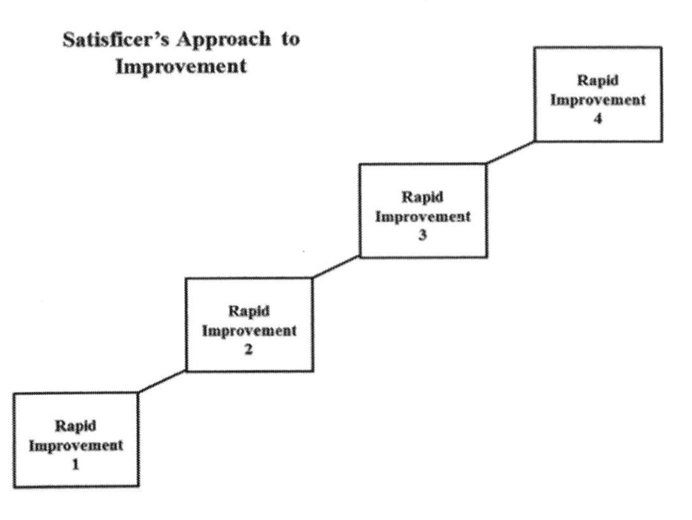

FIGURE 5.2
Satisficer's Approach to Improvement.

To effectively answer these five questions, as part of the Logical Thinking Process (LTP), Goldratt introduced a set of tools used to identify the root causes of negative symptoms, or Undesirable Effects (UDEs, pronounced oodees). These UDEs exist within virtually all organizations and business types. Goldratt believed that there are generally only a few core problems that create most of the UDEs we observe. And if we can identify these core problems (i.e. What to change?) and find their root causes and eliminate them, then most of the UDEs will disappear. Let's talk a bit more about these things called Undesirable Effects and how we can identify and understand them.

To understand what UDEs are, we must first understand that they must be considered in the context of an organization's Goals, Critical Success Factors, Necessary Conditions and Performance Metrics (PMs). For example, suppose the organization's goal is to make money now and in the future, and its Critical Success Factors and necessary conditions are things like generating enough revenue with low operating expenses, keeping its employees happy and secure, keeping customer satisfaction high, achieving superior quality and on-time delivery, etc. Further suppose that the organization measures its performance by things like on-time delivery, some kind of productivity measurement, the cost to produce products and a customer satisfaction index, and quality through parts per million (PPM) defective. Any organizational effect that moves the organization away from its goal, or violates one of the CSFs or NCs, or drives a PM in a negative direction with respect to its target, is considered undesirable. So, think for a minute about what UDEs might exist in your company.

The tool Goldratt developed to expose system type problems, or policy constraints, is referred to as the Current Reality Tree. The CRT is used to discover organizational problems, or UDEs, and then works backward to identify at least one root cause that leads to most of the UDEs. Dettmer [1] defines a root cause as, "the lowest cause in a chain of cause and effect, at which we have some capability to cause the break." His point is that the cause and effect chain could continue on indefinitely. But unless the cause lies within the span of control of the organization, or at least within its sphere of influence, it will not be solved.

Dettmer further explains that two characteristics apply to root causes:

1. It's the lowest point at which human intervention can change or break the cause.
2. It's within our capability to unilaterally control or to influence changes to the cause.

The CRT begins with identifying UDEs, or negative symptoms, existing within an organization, that let us know that a core problem exists. Core problems are unique in that, if the root cause or causes can be found, they can usually be traced to an exceptionally large percentage of the undesirable effects. Actually, Dettmer [1] suggests that this percentage could be as high as 70 percent and sometimes higher. Dettmer refers to a CRT as a "snapshot of reality as it exists at a particular moment in time." Dettmer further explains, "As with a photograph, it's not really reality itself, just a picture of reality, and like a photo, it encloses only what we choose to aim at through the camera's viewfinder."

By aiming our "logical camera" at the undesirable effects and their root causes, we're essentially eliminating all of the details that don't relate to them. In other words, the CRT helps us focus in on and pinpoint core problems. There are several different versions of the CRT available in the literature on the subject, but they all provide the same end product, which is at least one actionable core problem. Some CRTs are very detailed, while some are more general in nature.

The example we will present is a company that was having a problem generating enough Throughput (i.e. capacity constraint). They had plenty of orders, but were unable to produce enough parts to satisfy the market demand. It is clear to us that many of the problems that organizations encounter on a daily basis are really interconnected systems-related problems. It is further clear that by focusing on these core problems, organizations can essentially kill multiple birds with a few stones!

It is not my intention to present an in-depth discussion of Current Reality Trees, or how to construct one in this chapter, but I do want you to be aware of their existence. I will present a simple example that we developed for a company that produces electronic devices. This company had serious problems generating enough Throughput to satisfy the volume and delivery requirements of their customers. By creating a CRT, this company was able to pinpoint specific system problems that were constraining their Throughput and then take actions to alleviate the problem. I'll show you the step-by-step basics of how to create a Current Reality Tree and then expand upon how to use one in your company. Because a CRT is fairly detailed, I want to go through it slowly so that you can appreciate its usefulness.

THE CURRENT REALITY TREE

Before we begin to construct a CRT, we need to discuss something called the *Categories of Legitimate Reservation* (CLR), which will act as our "rules of engagement" for construction of Current Reality Trees. The Theory of Constraints Thinking Processes rely heavily upon the intuition of the individual using it. For this intuition to be useful, it usually must be verbalized and communicated to others. Since the verbalization process can be difficult for many people, TOC has developed some intuition-helping tools.

Dettmer [1] tells us that the CLRs are "the foundation upon which logic in general, and the LTP's in particular, are built." The CLRs serve to help solidify the logic of each causal connection. The CLRs also help us construct our own logical relationships and help us evaluate the logic of others. That is, the CLRs help us all avoid errors in logic as we progress through the construction of our Current Reality Tree.

The understanding and use of these tools is essential for constructing and verifying the validity of cause-effect-cause relationships. Once verbalized, these tools also play an important role in the communication process. There are eight different CLRs, with each serving a different purpose. And while they're not difficult to understand, they do require some practice using them, to keep them in your head. Or you can do what some people do and simply use a cheat sheet.

1. Clarity: Be certain that the individual words used in the various boxes are understood by everyone involved in the construction of the CRT and are a clear grasp of the idea being presented, and there is an obvious connection between the cause and the effect being introduced.
2. Entity Existence: Entities are complete ideas expressed as a statement. When constructing the graphic blocks (entities), be sure that the text is a complete sentence, not a compound sentence, and the idea contained in the sentence is valid and legitimate. Normally there is evidence to demonstrate its validity.
3. Causality Existence: The cause and effect relationships must really exist, and there should be no doubt in anyone's mind that "if we have this," "then we will definitely have that." A clear cause and effect relationship must exist.
4. Cause Insufficiency: Sometimes it requires more than one cause to be present to create the predicted effect, so be certain that you have identified and included all major contributing causes.
5. Additional Cause: It is possible that two completely different causes will result in the same effect, so each time you observe or imagine an effect, you must consider all of the possible independent causes.
6. Cause-Effect Reversal: Don't mistake an effect for a cause. People sometimes confuse the effect for the cause, so be careful.
7. Predicted Effect: This category is firmly rooted in the scientific method as evidenced by its primary function—to strengthen or break the proposed hypotheses. Predicted Effect may be used to test the validity of entities or causal relationships. It focuses the user on seeking the valid effects that must stem from the existence of the causality or the entity if they are valid. The Predicted Effect category asks the following question: Does another entity co-exist that will either strengthen the causality entity or disprove it? From a single cause can come many effects, so be sure to list all of the possible effects that you know about. This is where the team approach to CRTs becomes effective.
8. Tautology: This is sometimes referred to as circular logic because the effect is offered as a rationale for the existence of the cause. Don't take

the effect as unequivocal proof alone that the cause exists without considering other alternatives.

COMMUNICATING PRODUCTIVELY

When two or more people are having a discussion, how do they communicate differing perspectives or ideas? We have all experienced discussions that deteriorate into fruitless arguments. These discussions usually take some time to unravel; time for individuals to understand what is being proposed, and to determine if they can agree on a conclusion. Why does this happen? Often, it is because we don't know how to constructively scrutinize our claims and the claims of others. Usually this situation is magnified, because we also do not know how to communicate our concerns to others in a way that does not lead to defensive reactions.

When used to verify causality, the CLRs greatly diminish the impact of the first phenomenon (not knowing how to constructively scrutinize claims). CLRs can also be used in a specific order to promote non-defensive, focused, productive discussions. This process is based on four valid assumptions:

1. It is more effective to give people a chance to explain what they mean than to attack what we've understood them to say.
2. People are responsible for substantiating their claims.
3. People are not idiots.
4. What is said, and what is meant, are not always the same thing.

OK, now that you know about the CLRs, how do we use them, and what is their real purpose? Unlike the Goal Tree, which is based upon necessity-based logic, the CRT uses sufficiency-based logic. Whereas the Goal Tree was read as, "in order to have (x), I must have (y)," CRTs are read in an "if-then" form. So, to determine sufficiency, we might ask questions like "Is this enough to cause that?" or "Is this sufficient to result in that?" In short, a sufficiency tree implies that the causes are sufficient to actually produce the effect. As we construct the CRT, there will be more clarity on how we use the CLRs to construct them.

A Current Reality Tree is a logic-based structure designed to illustrate current reality as it actually exists now, or how it previously existed. As such, it reflects the intrinsic order of the cause-effect-cause phenomenon. Sometimes when you are faced with solving a problem, being able to precisely verbalize the problem is 95 percent of what is needed to solve a problem. From a preliminary problem analysis, you can quickly understand ALL the things that are going wrong (the effects), but you aren't sure why they happen (the cause). Sometimes, what might appear to be many different negative effects happening all at once can be reduced to a single core problem. If the core problem is removed from reality, then there is a high probability that ALL of the negative effects will be removed as well. Especially if the cause-effect-cause relationships can be established.

When faced with many different negative effects, the question becomes—"Which one do I focus my attention on?" When you review and analyze the negative effects, there are usually many things you can fix, but which one should you really focus your attention on? Which one provides the long lever that, if it is removed, most of the other problems will go away? The CRT provides such an analysis tool. The CRT is a powerful logical thinking tool that can help you filter the insignificant many from the important few. The CRT is probably not a document that you will assemble and analyze in 15 or 20 minutes. It will take some time, but the effort is well worth the results.

Current Reality Trees are *sufficiency-based logic structures* that enable individuals to investigate situations with a high degree of assurance that they are distinguishing reality from fiction. When people are asked to "find out what's going on" in a given situation, their first step usually involves gathering data. They take the data and categorize it, while looking for correlations. Often, the categories are prioritized according to the investigator's intuition about the existing correlations. This method of investigation is helpful, to a point. Classifying things to be dealt with or considered separately is not as efficient or effective as identifying one or two focus areas that will significantly impact the remaining areas in a positive way. By revealing and examining the underlying intrinsic order of entities, using *cause-effect-cause* relationships, one gains the ability to

1. Distinguish fact from fiction without spending a lot of time gathering data.
2. Focus on a core problem instead of multiple symptoms.
3. Succinctly communicate the past or current situation to others.

The Current Reality Tree is based on the fundamental natural law that order does exist. Events within a system do not happen with anywhere near the randomness that people think they do. Something happens (effect) because something else happened (cause). This technique provides the means for careful examination of hypotheses and assumptions and pursuing common causes that account for more and more of the effects in the system.

When using a Current Reality Tree to determine "*what to change*" in your existing system, you should search for those few entities that are causing most of the Undesirable Effects in your area of concern. It is always possible to build a comprehensive enough Current Reality Tree, in which at least one cause leads to the existence of most of the UDEs.

Figure 5.3 displays an example for the structure of a CRT. As a sufficiency-based logic tree, it is read using the "IF" and "THEN" statements. In other words, the logic is, if the entity at the base of the arrow, then the entity at the tip of the arrow. The arrow represents the logical connection and signifies that the entity at the base of the arrow is sufficient to cause the entity at the tip of the arrow to exist. As an example, IF the car battery is dead, THEN the car won't start. Obviously if the battery is dead, it is sufficient to keep the car from starting. At this stage, it is important to apply the Categories of Legitimate Reservation and validate the arrow's existence. In other words, make sure the arrow is logically solid.

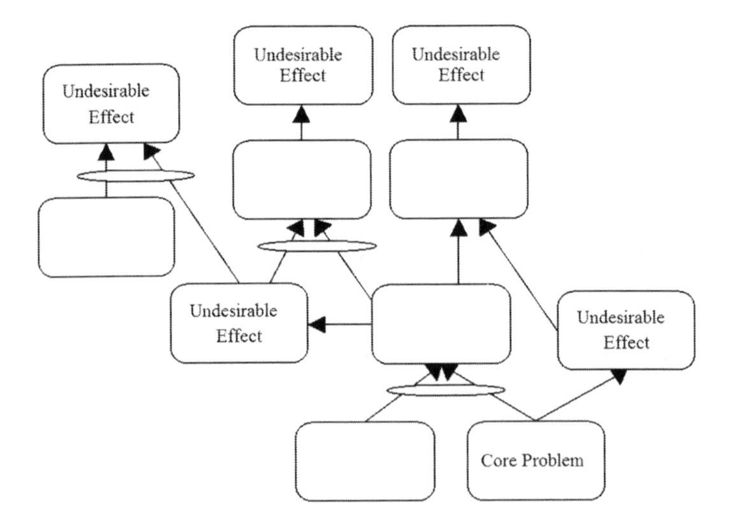

FIGURE 5.3
Example for the structure of a CRT.

In the example CRT, you notice ellipses that combine some of the arrows. These ellipses are the logical "and" statements. So, reading the statement becomes, "If entity #1 'AND' entity #2, THEN entity #3." The "and" statement implies that both entities (causes) are required to generate the effect.

Current Reality Trees can be used to

1. Identify the core problem—What is constraining the Throughput of the system?
2. Focus and leverage improvement efforts.
3. Determine what is happening or what has happened.
4. Communicate information about a past or a current situation clearly and concisely to others.
5. Provide a pattern from which future events may be predicted.
6. Analyze the validity of an article or argument.

STEPS TO CONSTRUCT A CRT

1. **Choose the topic you wish to address.** Do you want to find the weakest link in your area of responsibility? Do you want to investigate why certain things happen? Do you want to understand your teenager? This step results in concisely defining the system you want to analyze.
2. **List the Undesirable Effects within the chosen system.** Focus and direction are gained through clarifying the preliminary boundaries of the area to be analyzed. Undesirable Effects should relate to the unsuccessful attainment of the goal (or Necessary Conditions) as revealed by its measures. You can find the UDEs by asking the question, "When I think about this system it bothers me that ..." Your list of "it bothers me that ..." statements translates into the list of UDEs.
3. **Map out the causal connections among the Undesirable Effects.** Through mapping the sufficiency relationships that connect the UDEs, common causes begin to take shape, and the picture of the current situation becomes clearer. This step involves rigorous use of the Categories of Legitimate Reservation (logic-checking tools) to minimize self-deception. What you are looking for is a relationship

between the UDEs. Is there a particular UDE that exists because another UDE exists? If there is, draw the arrow between your two entities, and apply the CLRs to check the validity.

4. **Modify the tree to reflect your intuition about the system being analyzed.** The previous action had you working at a micro-level. This action provides a safety net—just in case you "can't see the forest for the trees." You may find this step enables you to limit or expand the analysis so that it is in proper perspective.
5. **Identify those entities you perceive to be the most undesirable.** Your initial list of UDEs was a mini brainstorming session. At this stage of the game, you should have a much clearer picture of what is going on and a better sense of what the important UDEs are—the ones that are impeding the organization.
6. **Trim entities that do not participate in connecting the major Undesirable Effects.** When seeking to find the constraint in a given system, one must focus on the connections that matter—the connection between a common cause and the effects that you want to eliminate. This step asks you to remove the extraneous entities. The point here is to trim the UDEs that fall outside your area of control or sphere of influence. It helps establish the focus and defines the boundaries of what you really want to analyze. Knowing this information provides the *leverage* needed to accomplish the effort.
7. **Identify the core problem.** This final action entails examining the entry points (entry points are UDEs that DO NOT have an arrow pointing to them) to the tree and finding one that is responsible for a significant portion of the major UDEs. Which entity is responsible for most of the UDEs? A general guideline is whatever entity is causing 80 percent of the UDEs is usually a good candidate for the core problem.

The CRT allows you to focus on those one or two things that can really make a difference and concentrate your resources on solving that particular problem. In other words, you gain the ability to filter the insignificant many from the important few. When you focus on the proper core problem and remove it, the other UDEs, by default, will also disappear. If the causality is gone, so is the negative effect. So, now you have found the core problem—now what? You've found "What to change." Now we need to find "What to change to," or answer the question "What is the best solution to solve the core problem?"

We will now create a simple, but real, CRT to demonstrate what a completed CRT looks like. In the last section, we discussed the elements of the CRT, with the promise to share a CRT from a real situation. Without revealing company names, let's start with some background on this particular company.

This company is a major producer of electronic components, mostly in the form of circuit cards. They are major supplier to other companies in the electronics industry. The plant was configured with seven major assembly lines. Most lines were dedicated to certain types of boards, but there was also several with cross-functionality. In other words, the same type of board could be produced on more than one line. The most notable problem, and the reason they called us, was they were suffering from very high levels of work-in-process (WIP) and not being able to meet on-time delivery demands from the customer. We started our analysis with them by interviewing the workers on the line. We were first looking for the perceived UDEs that existed.

COLLECTING UDES

The UDEs provide a very important piece of the puzzle you are trying to put together. But, beware—not all UDEs are really UDEs. It's important when you collect UDEs to have people write them down in the form of answering a question. For example, you might ask, "When I think of the current system, it bothers me that …" The "that statement" becomes the UDE. The more people you talk with, the better and more refined the UDE list will become.

Another important factor is to note the commonality between statements. Five or six different people might all say something different, but all six might mean exactly the same thing. When you find a UDE that fits this category, you've found an important UDE. It is also important to filter the UDEs, to separate those emotional statements from logical statements. As an example, suppose during the UDE collection, someone responds to the statement with "It bothers me that my boss is an idiot!" No matter how true that statement may or may not be, it is an emotional statement and not a logical statement. Spending the necessary time on the front end to filter the UDEs can translate into a much smoother process when constructing a CRT.

With that said, here is the final UDE list developed for this company:

1. The front of the line is measured in utilization minutes.
2. The back of the line is measured in boards per day.
3. Raw materials are sometimes not available for production runs.
4. Testing takes too long to complete for some boards.
5. FTC, a type of Manufacturing Engineer, and Customer Quality Assurance (CQA) perform the same function.
6. Some test equipment is not effectively used.
7. 100 percent of the boards are tested.
8. Boards can be rejected for cosmetic reason and not functionality.
9. Some batch sizes for some boards are too large.
10. Some Finished Goods (FG) sit in testing, waiting for transfer to FG inventory.
11. Testing is not considered part of the production line.

From the 30 or so different UDEs collected, the list was reduced to the previous list. Each UDE seems to be a separate problem, with no clear correlation between them, and each is causing its fair share of Undesirable Effects in the system. So, the hunt was on to discover correlation between the UDEs and surface a probable Root Cause.

CONSTRUCTING THE CRT

With the UDE list, we are trying to build correlation between the entities. In other words, are there any two of these entities where one can cause the other? When you find those two, it becomes the starting point to build the rest of the CRT. Continue building until all, or most, of the UDEs, have been connected. Figure 5.4 shows how these entities were connected to create the CRT. You'll notice that each entity box contains a number at the top. This is nothing more than an entity address that helps when scrutinizing using the CLRs, to be able to point out entities quickly so that the connection can be made. Those entity numbers with an asterisk (*) were entities from the original list, while those without an asterisk are the entities that surfaced during development of the CRT as predicted effects and additional causes from the CLRs.

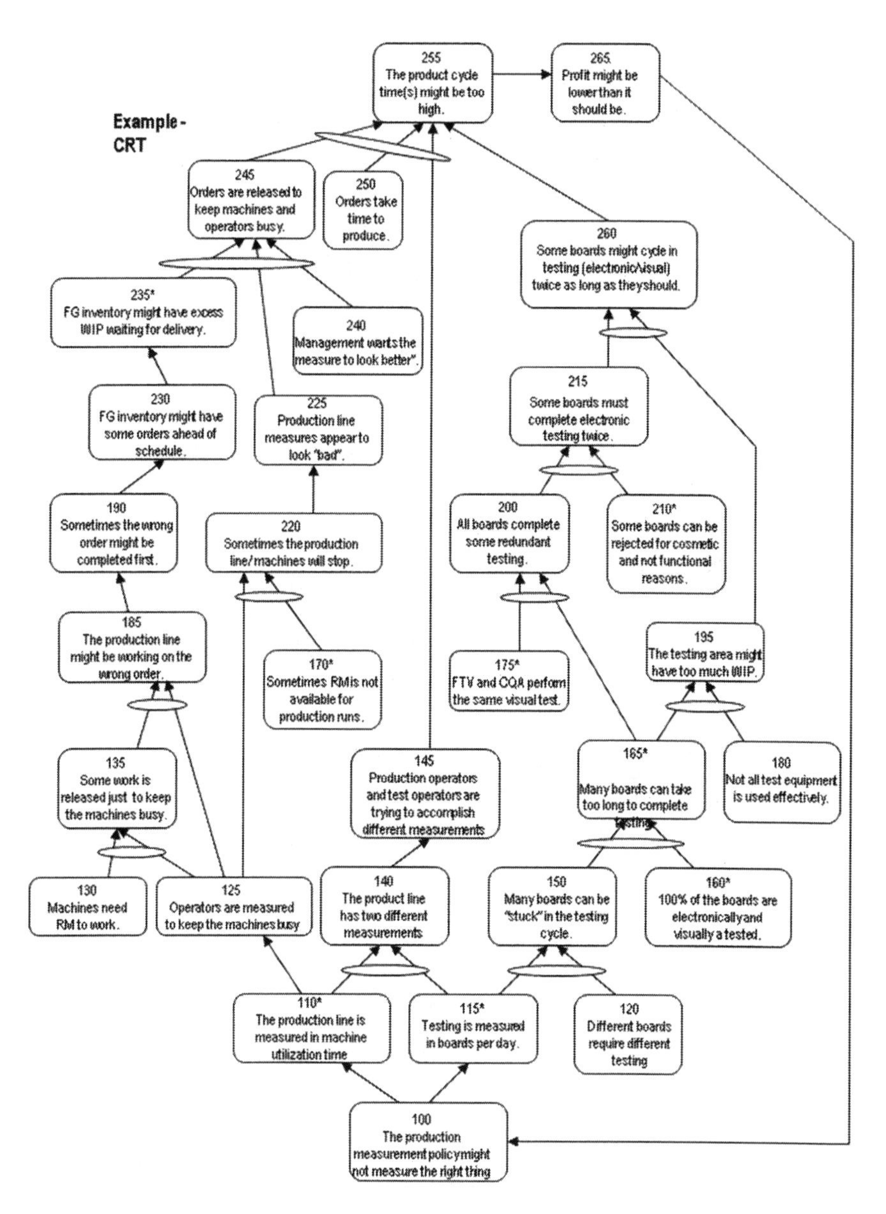

FIGURE 5.4
How UDEs Were Connected to Create CRT.

Using the CRT to formulate the sufficiency-based logic, you can see that from the original UDE list, we were able to show cause-effect-cause relationships between all of the undesirable effects. The root cause in this example WAS NOT an entity listed in the original list, but rather a root cause that was exposed because of the CRT. In this case, it was policy constraints. I say constraints as plural, because this company has so many measurements they were trying to record, and some of these measurements were in direct conflict with each other. You'll notice at the bottom of the CRT, the two measurements, one for minutes and one for boards. In their mind, high machine utilization was equal to producing lots of boards. They had very expensive equipment, and the only way they could justify the equipment was to keep it busy all the time. Because of this measure, they continually loaded the system with work, which created vast amounts of work-in-process inventory and much longer lead times and, consequently, missed due dates.

In the end, policy measurements were changed or eliminated, and the system was structured as a Drum Buffer Rope system, with the test equipment being the drum. By using the test equipment as the drum, we were able to release ("pull") work into the system at the correct rate. This was a much different environment than trying to "push" work into the system, for the sake of efficiency. The overall WIP reduced dramatically, the lead times were shortened to hours rather than days, and on-time delivery skyrocketed. Revenue jumped $350M in six months' time, and all because a CRT helped them understand what the real root cause was. In our next section, we'll look at the Conflict Diagram (CD) and review some basic principles for its use and structure.

CONFLICT DIAGRAMS—BASIC PRINCIPLES

Of all the tools associated with the Thinking Processes, we have found that the most useful, and the one we use most often, is the Conflict Diagram. It seems that solving conflicts is part of everyday life. No matter how much you plan, no matter how careful you are, there always seems to be conflicts. So, if conflicts are a normal aspect of everyday life, and decisions are required to solve conflicts, then having access to the best tools to resolve conflicts is paramount.

Conflict Diagrams are necessary condition-based (necessity-based logic) structures, used to verbalize and resolve conflicts (dilemmas). In the past, this thinking tool has come to be known by many different names. In the early days of TOC development, it was known as the "evaporating cloud." It has also been referred to as the "Conflict Diagram" and the "Conflict

Resolution Diagram". Whichever name you choose, the structure remains the same. We have chosen the term Conflict Diagram.

Typically, resolving conflicts involves investing time (sometimes large quantities of time) in finding a compromise on which both sides will agree—however reluctantly. Yet there are many times when there is no acceptable compromise that both parties will agree to. The problem with a compromise is that both sides have to give up something to achieve common ground. When a compromise is used, the end result is usually so diluted that it jeopardizes the achievement of an important objective. Unfortunately, many objectives are compromised through this process of seeking consensus on a solution. And in the end, the results are not satisfying for either side. The compromise process usually results in a "lose-lose" situation. In other words, neither side achieved what they really wanted. If such is the case, wouldn't it make more sense to spend the necessary time trying to eliminate some conditions (assumptions) in reality, that need changing, rather than to compromise the objective? In the process of compromise, it makes sense that breakthrough ideas are usually hidden from us because we are geared to looking for compromises. Perhaps generating the idea (injection) that creates a "win-win" solution, without a compromise, would be a much more acceptable platform to resolve conflicts.

By rejecting the tendency to compromise the stated objective, one gains the ability to

1. Set objectives based on what is wanted or needed, rather than on that which is currently deemed possible.
2. Challenge vital assumptions that sustain the conflict.
3. Find paradigm-shifting ideas that increase the likelihood of achieving the objective.

The Conflict Diagram provides a concise verbalization of a problem, but what is a problem? It is usually defined as a situation where you are unable to get what you want. From this definition, it is easy to see that one element of the Conflict Diagram is a description of what you want, the objective, and another element of the Conflict Diagram is a description of something that is preventing the achievement of the objective. In essence, we are clearly defining the conflict. Once the situation is clearly defined and the entities of the conflict are clearly verbalized, the stage is set for generating breakthrough ideas.

Most people have had an idea come to them from out of the blue. Maybe you have been thinking about a problem, and then all of the sudden you wake up one morning and a brilliantly simplistic solution comes to you. What blocked you from being able to solve the problem before? You probably had some assumptions about a necessary condition that didn't really have to be necessary. Once you realized this at some level, you were able to come up with a way out of the predicament. Finding breakthrough ideas comes through challenging assumptions we make about our reality. The assumption-based thinking (human behavior) is an essential part of the Conflict Diagram. Figure 5.5 is the basic structure of the Conflict Diagram.

In this diagram, entity "A" is the Objective, which is the statement that defines what you really want to do. The "B" requirement is a statement that defines something that must exist to achieve the Objective "A." The "C" requirement is the statement of an additional requirement that must exist to achieve the Objective "A." "D" is the prerequisite statement (entity) for "B" and "E," which is the prerequisite statement for entity "C." The conflict when it is surfaced will reside between "D" and "E" (hence, the lighting-bolt arrow). The statements written in "D" and "E" will usually be opposite statements. For example, the statement in "D" might say "do something," and the statement in "E" will say "don't do something." The line between "D" and "E" represents the tug-of-war between the two statements. As a necessity-based structure it is read, "In order to have 'A,' I must have 'B.' In order to have 'B' I must have 'D.'" The same rules apply to the lower leg of the diagram: "In order to have 'A,' I must have 'C,' and in order to have 'C,' I must have 'E.'"

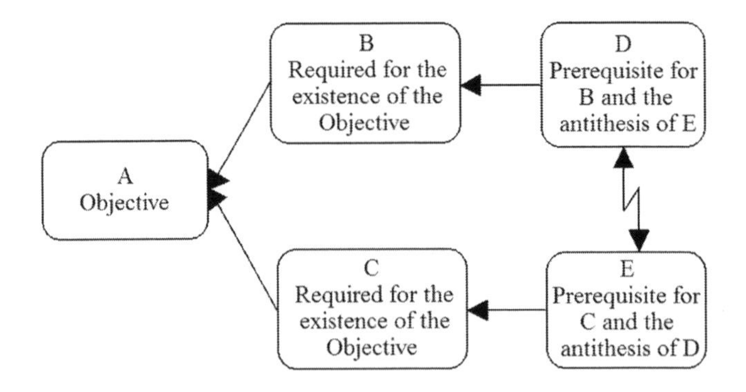

FIGURE 5.5
Basic Structure of the Conflict Diagram.

The structure of a Conflict Diagram is such that it presents both sides of an argument or dilemma. Each side of the dilemma is commonly referred to as a leg. One leg presents your side of the conflict, while the other leg represents the other side of the conflict. Each leg appears valid to the person or group presenting those statements. Each leg appears to define the "necessary conditions" to achieve the objective. But each leg is in conflict with the other leg. One side says, "Do this," and the other side says, "Do that." You can't do both, so what do you do? In the next section, we will discuss the uses for Conflict Diagrams and the steps to construct a Conflict Diagram.

CONFLICT DIAGRAMS—STEPS TO CONSTRUCT AND CATEGORIES

In the last section, we discussed some basic principles for the Conflict Diagram. In this section, we will discuss the steps to construct the CD and the different categories that a CD might fall into.

The Steps

There are basically four steps to construct a workable CD, and following them in order presents the best chance for success.

1. Identify a scenario where you know, or suspect, a conflict is blocking the achievement of an objective.

   In today's ever-changing and demanding world, there always seem to be many situations in which we think there is "no way out." The feeling is that we are somehow caught between a rock and a hard place, with no way to resolve the problem. This feeling of frustration occurs only when the conflict is blocking the pathway for achieving something that we really want. These types of dilemmas are the situations that benefit the use of the Conflict Diagram.
2. Concisely verbalize the dilemma.

   Sometimes, the real dilemma or problem in a situation is not as clear as it should be. By concisely verbalizing the dilemma, you are halfway toward solving it. If you don't clearly understand the problem, then it is very difficult to provide an effective solution. By identifying the conflict objective, requirements and prerequisites that make up the dilemma, you will develop a very concise verbalization of the dilemma. Verbalizing and understanding the dilemma provides an excellent starting point for effective problem solving.

3. Surface the assumptions that support the existence of the dilemma.

   Within whatever system you are operating, there are many things that we assume to be absolutes. In other words, they appear to be things we assume we can't change. There are many instances we assume these things to be concrete, things that are somehow unchangeable. Yet, there really are very few things in systems that are unchangeable. The purpose of Step 3 is to separate the wheat from the chaff. In other words, a way to separate the actual "facts" from the perceived "fiction." Once you understand what's real and what isn't, the task becomes much easier.

   The assumptions are the answer to the question on the solid lines of the diagram. In other words, "In order to have A, I must have B, because …?" It's the assumptions that make the lines solid, and we want to break an assumption. When the assumption is broken, the line is no longer solid, and the conflict can be resolved.

4. Generate breakthrough ideas that will invalidate at least one of the assumptions.

   Generating the breakthrough idea is the primary function of the CD—that is, to surface an idea that completely eliminates the assumption. When you surface the assumptions on each arrow, ask yourself, "What must exist in order for the assumption not to be true?" Whatever the statement is, this is your injection or idea to replace the assumption.

CATEGORIES OF CD

Most Conflict Diagrams will fall into one of four categories. This section provides some hints about how to construct the CD, based on what type of category it falls into. These definitions will tell you which entity to fill in first, which one second, etc.; where you should start; what you should do next; and what steps to take within each of the categories of CDs. The following are examples of each of the different categories of Conflict Diagrams.

1. Negotiation (Figure 5.6): Your opinion versus someone else's.
2. Crisis (Figure 5.7): Current reality versus desired reality.
3. Classic (Figure 5.8): I know what I don't want.
4. Core Problem (Figure 5.9): Concisely verbalizing the core problem from a CRT.

By constructing and using the CD, you have surfaced the injection or idea that you want to implement into your reality. How do you know if your idea will generate the results that you want or need? In the next segment, we will discuss the Future Reality Tree (FRT) as a means of testing ideas, not only of the positive impact but also to surface any negative branches that might exist with your idea.

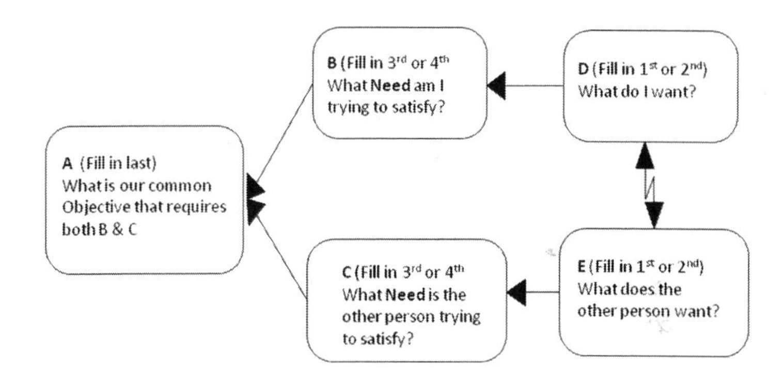

FIGURE 5.6
Negotiation: Your Opinion versus Someone Else's.

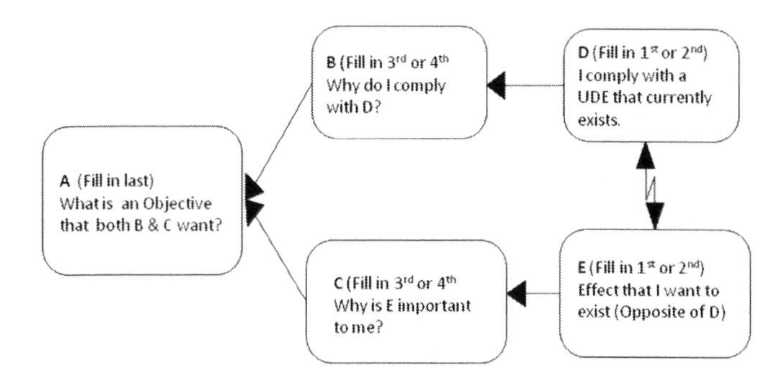

FIGURE 5.7
Current Reality versus Desired Reality.

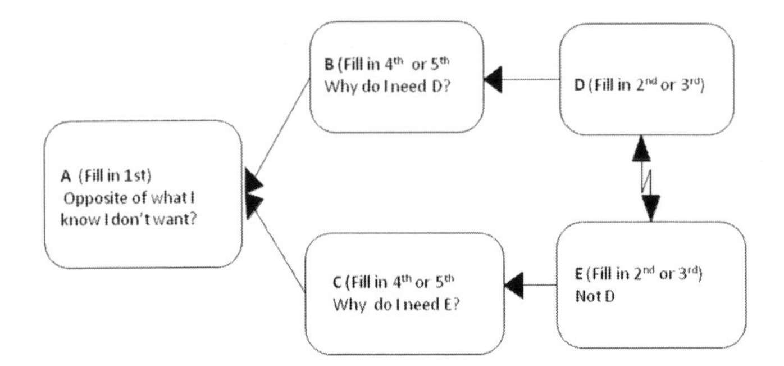

FIGURE 5.8
Classic: I Know What I Don't Want.

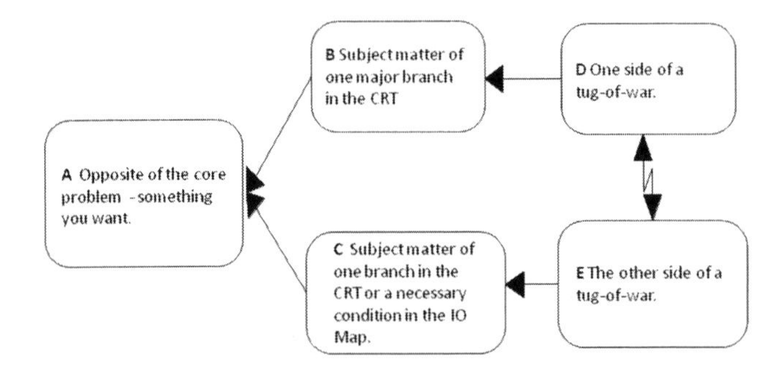

FIGURE 5.9
Concisely Verbalizing the Core Problem from a CRT.

FUTURE REALITY TREES—BASIC PRINCIPLES

In the last section, we discussed using the Conflict Diagram to break the assumptions of a conflict. When you break an assumption, you create an injection or idea. The injection is something that, if it existed, would mean the assumption would be broken, and the conflict would be resolved. It is possible to generate several different ideas, each sufficient to resolve the conflict. The choice now becomes, which injections do you want to pursue, or which one gives you the best results? We use the FRT to test these ideas.

A Future Reality Tree is a sufficiency-based logic structure that is used to check ideas. It is possible that an idea that has good possibilities for success might also contain some flaws or negative effects. Knowing that some bad can co-exist with the good is probably why the common technique used to evaluate an idea/decision is a list of pros and cons. This technique isn't all bad, but unfortunately, it does not provide enough information as a means of systematically elevating your idea to create a good solution. Nor does it enable you to check if the "pros" will really result from the idea. The FRT first validates that the selected idea will lead to the desired results. If it doesn't, the idea is supplemented with additional injections, until all desired results are achieved. Then, the FRT uses any potential problems (negative effects) of an idea as a means to improve the idea instead of a reason for dismissing it.

The Future Reality Tree is based on three fundamental assumptions:

1. It is better to know what the idea yields, before acting on it.
2. The future is predictable, to the extent that current causalities are understood.
3. Negative side effects, as long as they are determined before the idea is implemented, provide the means for improving the idea.

In many ways, the Future Reality Tree is a simulation model for ideas. It simulates the system to react to an idea with no limits on the number of ideas that can be simultaneously tested. Using the FRT, the existence of injections (ideas) is assumed, and their inevitable effects are predictable, using well-scrutinized cause-effect-cause relationships. Usually, single injections are insufficient to cause the desired effects, but during the process of building the FRT, additional injections can be discovered that are needed to reach the desired result.

Sometimes a brilliant idea can turn sour. Has it ever happened that what seemed like a good idea produced less than the anticipated results? That what seemed to be a good idea in the beginning quickly starts to generate some negative effects? The old adage that tells us, "too many times the medicine is more harmful than the disease" could very well true.

Remember, as a sufficiency-based structure the tree is read "If the base of the arrow, then the tip of the arrow," and the ellipse represents a logical "and" statement. The additional injections noted in the example tree are the ideas you have generated along the way to keep your good idea on track and stable—those additional things that must exist for your idea to work.

FUTURE REALITY TREE USES

Future Reality Trees can be used to

- Test the merits of ideas, before taking action.
- Construct a solution that yields a high degree of assurance that the existing undesirable effects will be eliminated without creating devastating new ones.
- Check for and prevent potential negative ramifications of an idea.
- Build a strategic plan.
- Verbalize and communicate a vision.

An FRT can take some time to construct, but a good FRT is worth the effort. It is always better to test your ideas before implementation than to find out after the fact that the idea wasn't so good after all. It's a way, if you will, to view your idea in "fast forward" and make sure you like the end results. If you don't like the results, and you can't come up with additional injections to nullify the negative effects, then go back and select another idea to implement. Figure 5.10 is the basic structure of the FRT.

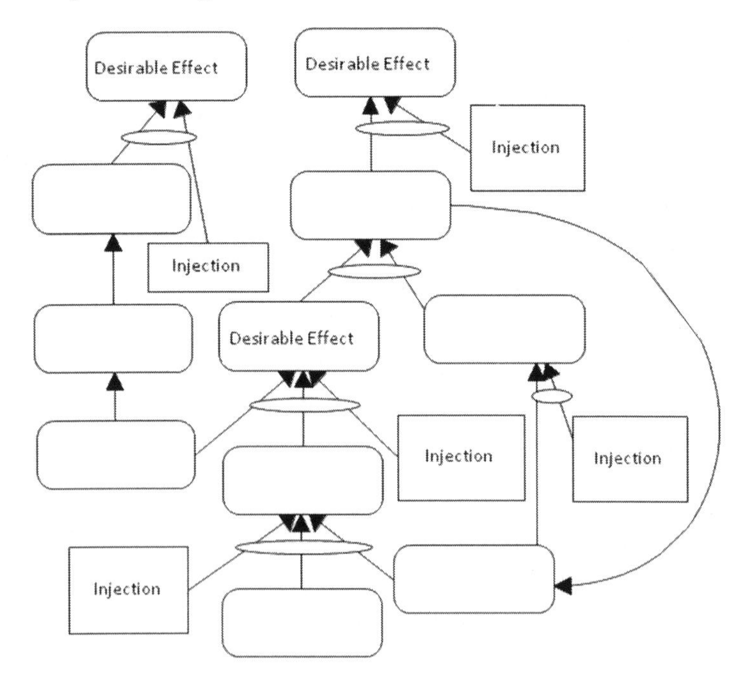

FIGURE 5.10
The Basic Structure of the FRT.

FUTURE REALITY TREES—STEPS TO CONSTRUCT

In the last section, we discussed the basic principles and uses of the Future Reality Tree. In this section, we will discuss the necessary steps on how to construct an FRT.

The following steps provide the intrinsic order of thinking when constructing an FRT. It is best to follow the steps in the order given to construct the best possible tree.

Step 1—Define the Function of the Future Reality Tree

When constructing an FRT, it is always best to define the function. In other words, "why are you constructing the tree?" A Future Reality Tree can be used to construct a full solution that you want to implement. It can also be used to test an idea (yours or someone else's), or to present the merits of an idea to someone else. When you focus on what role you want this technique to be used for, then the remaining analysis will be more relevant.

Step 2—Capture the Idea

There can be a significant difference between a "new" idea and a "good" idea. A good idea is one that accomplishes its objectives, without creating unwanted negative effects. In this action, the new idea is captured verbally. Again, capture your idea as succinctly and concisely as possible, a single statement that captures clearly what it is you really want to do.

Step 3—Make a List of Potential Desirable and Negative Effects

One thing you want to verbally capture is the desired positive effects that you would like to see happen. The list of desired effects will depend on the type of FRT you are constructing, Suppose you are constructing an FRT to complement a full TOC analysis, then you can use the Undesirable Effects from the CRT as a guide. For example, suppose one of the UDEs from the CRT was "ROI is too low," then the desired effect would read "ROI is high." Continue building your Desired Effects list until you have listed all, or most, of the things you want.

There may be some expectations, as well as possible concerns, about what this implemented idea will accomplish. What you are looking for are the "good" things that will exist when the idea is implemented. Write down the potential positive effects and the potential negative effects (Step 6) that this idea, once implemented, will cause. Be honest and logical, and spend the time necessary to filter those emotional statements from the logical statements.

Step 4—Build the Causal Connections between the Injection and Desirable Effects

You are looking for the causal relationship between two of your desirable effects. Can you see a connection between any two entities, where one would be sufficient to cause the other? If so, make the connection. If you are using the tree to validate an injection from a CD, then it is desirable for the injection to be near the bottom of the tree. What you are looking for are all of the desired effects that will come from the injection. At this stage, continue to look for and connect the other causal links between the injection and the desirable effects. In the course of constructing the tree, it is very possible you will surface additional desirable effects not on your original list. It is also possible that you will add additional injections to take care of the potential negative effects.

Step 5—Strengthen Your Analysis

A powerful outcome from constructing an FRT is to look for the positive reinforcing loops. In other words, those things that through time, just continue to keep happening. These positive loops help ensure that the solution will work the way you want it to, over and over again.

Apply the Categories of Legitimate Reservation (CLRs) and strengthen the logic. Are there additional positive effects (Predicted Effects)? Is there additional cause required to make something happen (Injection)? Full scrutiny with the CLRs will result in a powerful and useful FRT.

Step 6—Actively Seek Negative Branches

This is probably one of the most important aspects and outcomes from an FRT. That is, looking for the Negative Branches or Negative Effects from your idea. Don't be frustrated and think that your idea won't work because you found a negative effect. Quite the opposite is true. It now provides an opportunity to know that a negative effect is possible and allows the chance to inject with an additional idea to keep the negative effect from happening. It's much better to attack it now and have a way to overcome, rather than wait until it is actually implemented. It's part of the necessary planning. Applying the Negative Branch technique helps ensure that the medicine is not worse than the disease. If you don't find and resolve the negative effects as part of your solution, the negative effects will appear in reality and be much more difficult to solve.

Once you have completed the FRT and discovered the best idea that you want to move forward with, the next question becomes "What stops me from doing this right now?" In the next section, we will discuss the Prerequisite Tree (PRT). The PRT is used to define and overcome the obstacles that seem to be stopping the implementation of your good idea.

PREREQUISITE TREES—BASIC PRINCIPLES

In the previous section, we discussed the elements of the Future Reality Tree (FRT). With the FRT, you determined an injection or idea that you want to move forward with. With the FRT, you want to determine what are those obstacles that stop you from doing this right now. Many people might be inclined to offer reasons for "why" it won't work.

Prerequisite Trees are based on necessary conditions that provide the process to systematically dissect any major tasks into a set of smaller segments of more achievable intermediate objectives (IO). Each IO is determined as a necessary condition to overcome previously known, or perceived, obstacles. Once they are identified, the IOs are sequenced in the intrinsic order to accommodate for the existing time dependencies that will exist between them. The completed Prerequisite Tree presents the time sequence of the IOs and the stated obstacle(s) each is intended to overcome.

Whenever you try to implement change, it seems that the most frequent response is, "It won't work here because...." These "because" statements are often followed by an explanation of what are perceived to be the obstacles (sometimes many), which can delay, obstruct or completely block the achievement of the objective. In the majority of cases, the presenter neither actively seeks, nor greatly appreciates, the input of the naysayers.

However, when building a Prerequisite Tree, such input for obstacles is actively required. By surfacing the obstacles in advance, the implementer has the opportunity to plan strategy to overcome them, instead of waiting for them to block progress in reality.

Once the obstacles are identified, you need to create a specific IO, sufficient to overcome or eliminate the impact of the obstacle. Each IO, when achieved, must be sufficient to overcome one or more of the obstacles that block progress. When all IOs are achieved, the path to completing the objective is much more straightforward.

Sometimes, when you are assigned a major new project, the mere thought and scope of the effort can be daunting. It is difficult to figure out where to start and what to do. This difficulty is compounded even further by the fact that there could be many required IOs necessary to reach the stated objective. By defining the obstacles in your specific situation and determining the needed IOs, you can map the logical and intrinsic flow, or steps you must go through to achieve the desired objective. Often, you will find that simply defining and listing the obstacles to your objective will make it seem much more achievable to you and to others. In many cases, you will find the mystery has now dissolved. The Prerequisite Tree is a logical tool designed to drastically simplify organizing data for a large task. The intrinsic order of task completion will become obvious and set the foundation for a clearly defined implementation plan. If the logic is solid, then the implementation will be solid.

THE PREREQUISITE TREE (PRT)

Prerequisite Trees can be used to

- Set Intermediate Objectives for implementation of the solution.
- Systematically dissect a major task into a set of interdependent bite-sized pieces.
- Identify and overcome obstacles.

Figure 5.11 provides an example of the PRT structure. Note, on the example, that the objective boxes are square-cornered, meaning they are the Intermediate Objectives, or something that does not yet exist in

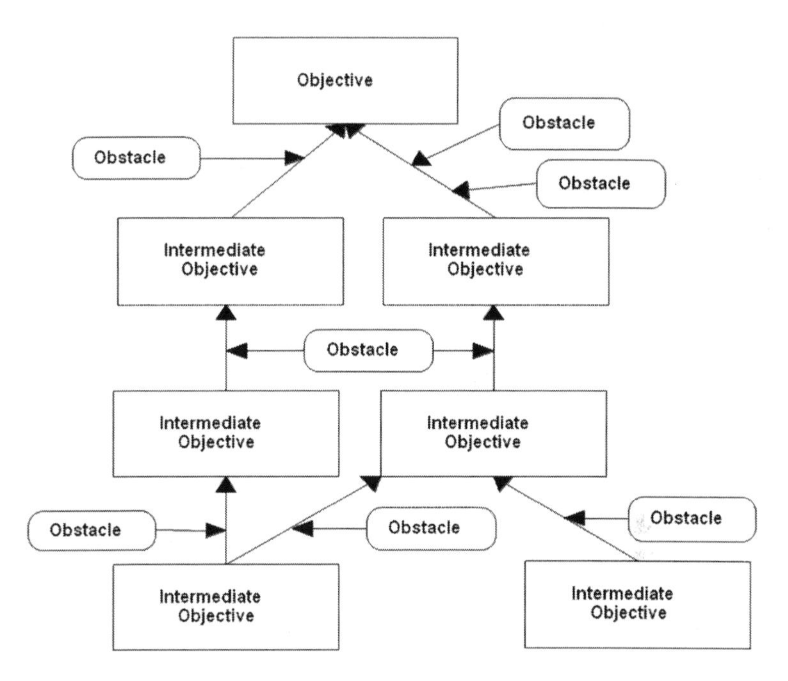

FIGURE 5.11
Example of the PRT Structure.

reality. The round-cornered boxes are the obstacles, or statements from reality, that do exist.

The tree is read, "In order to have ..." (entity at the end of the arrow) "I must have ..." (entity at the base of the arrow). This statement sets up the necessity for the arrow. The obstacles are pointing to the necessity arrow. So, when reading the entire statement, it reads, "In order to have ..." (entity at the tip of the arrow) "I must have ..." (entity at the base of the arrow), "Because ..." (the statement from the Obstacle entity). In the next section, we will discuss the steps to construct a PRT as well as some useful hints to consider.

PREREQUISITE TREES—STEPS TO CONSTRUCT

In the previous section, we discussed the basic principles that make up the Prerequisite Tree. In this section, we will discuss the steps required to construct a PRT and some helpful hints for each step.

Step 1—Verbalize Your Objective

A clear and complete verbalization of your stated objective will better enable you to stay focused. Without the necessary focus, there could be a tendency to wander off the path. The correct focus will help you achieve the ultimate objective. The best objective is to choose something that you truly want and which is beneficial to the system. It is possible when doing a full TOC analysis that the Objective could be an injection from your FRT, something you really want or need to benefit the system but aren't exactly sure how to achieve it.

- Identify a situation you have tasked to accomplish, and you sense it will be difficult. If you don't already know how to accomplish the task, this will help you define the necessary steps.
- Ask yourself what the purpose or objective of the task is. It must be something that we really want and is worth the effort to work toward achieving.
- State the purpose as a specific objective in the present tense. When using a Prerequisite Tree following a Future Reality Tree, choose as an objective the injection that you feel will be the most difficult (often the largest one) to achieve. Often, some of the other injections on the Future Reality Tree will be prerequisites to that difficult injection.
- Determine whether or not to use the Prerequisite Tree. Not every situation will benefit from the time and energy required to do a PRT. When you are using a PRT following an FRT, usually some of your injections must be broken down further using a PRT.

Hint: Look at the big picture. Ask yourself whether or not the objective(s) you have chosen present any major obstacles that you do not already know how to overcome.

Step 2—List the Obstacles That Prevent the Attainment or Existence of the Objective

Capturing all of the things that may block you from achieving your objective enables you to address each obstacle individually. People are very good at listing the reasons something "can't" be done. Typically, you will feel much better about your ability to reach the objective simply by surfacing the obstacles.

- Write down an Objective at the top of the page.
- If you are constructing the PRT following an FRT, begin with the injection that looks like it is the most difficult.
- In one column, write down the major obstacles you think stand in the way of achieving the objective.
- Check each obstacle, and check that you have written an obstacle to your stated objective.
    a. Check what you have written for entity existence. "Does the obstacle exist in my current reality?"
    b. Check what you have written for causality existence. "If the obstacle exists, then I will not be able to achieve the objective." It is possible you have not captured the true obstacle that is preventing reaching the objective. Be careful to keep the objective foremost in your mind so that you will not stray into obstacles that are not related to your objective. (FOCUS)

Hint: If you have two objectives/injections from the FRT work them separately. Start with the injection you believe to be the most difficult to achieve. You may find that subsequent objectives are actually IOs to the major objective you started with.

Step 3—Determine Intermediate Objectives That Eliminate the Obstacles You Have Listed

Tackling each obstacle individually helps to break the Objective down into a series of smaller pieces or Intermediate Objectives. Each Intermediate Objective should be sufficient to overcome its corresponding obstacle, and it should be more feasible for you to achieve than the Objective.

- For each obstacle on your list, ask yourself what would overcome it. At this point you are not necessarily trying to define the actions that you must take to achieve the objective, but rather to state the other things that you must accomplish on the way to it.
- Write down your idea as an entity in the present tense.

This entity is called an "Intermediate Objective". When doing the PRT after an FRT, you can use other Injections as the IOs (Figure 5.12). Sometimes one IO will overcome more than one Obstacle on your list. This is perfectly acceptable and can reduce the number of IOs required

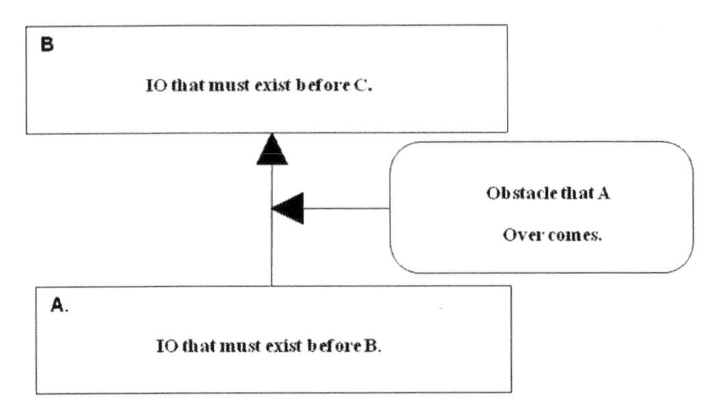

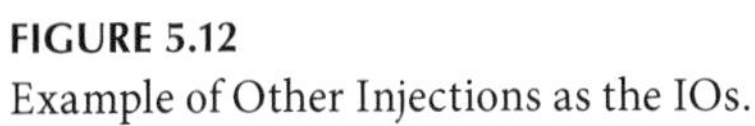
FIGURE 5.12
Example of Other Injections as the IOs.

to reach your Objective. However, each IO should be able to overcome the obstacle by itself. The IO should be more feasible than the Objective. Since we are trying to make the task of reaching the objective easier, each IO must be, in itself more manageable than the ultimate Objective. If it is not, you should search for something else that will eliminate the obstacle and be easier to attain.

Hint: If you have difficulty coming up with an acceptable IO, use the Conflict Diagram to generate more ideas. If you feel "stuck," it is usually because of some conflict that blocks you from overcoming the obstacle. The Conflict Diagram will help you expose the conflict and enable you to break it.

Step 4—Find the Time Dependencies between the Intermediate Objectives

Most of the time you will find time dependencies that exist between IOs, such that you cannot accomplish one without first accomplishing the other. These time dependencies establish the intrinsic order in which you must accomplish the IOs and work toward the objective. In essence, it provides the step for the implementation planning. Which one do you do first? Which one second?

- Identify two Intermediate Objectives that have an apparent time dependency between them. One must be completed before the other can be happen.

- Illustrate the connection. Necessary condition arrows are between the IOs, Sufficiency arrows are from the obstacle to the necessary condition arrows.
- Scrutinize the connection. "In order to have (IO at the tip of the arrow), I must have (IO at the base of the arrow), because of (Obstacle at the base of the Sufficiency arrow)." AND "I cannot have (IO at the tip of the arrow) because of (Obstacle at the base of the Sufficiency arrow)." It may be necessary to add other IOs and obstacles from your list to bridge and validate the connection. Sometimes there is a time dependency between two IOs, but it is not a direct one. In these cases, you will need to place other IOs in between your original connection to make it more intrinsically logical.
- Connect additional IOs from your list to this original cluster. Try placing the other IOs in their appropriate time dependencies sequence with the first cluster. Scrutinize each connection you make as in Step 3. If an IO doesn't seem to fit, do not connect it to the others. This means it probably does not have a time dependency with the other IOs and can be achieved without first accomplishing other IOs. You may have more than one grouping in a PRT, as well as some IOs that don't seem to fit anywhere.
- Connect the IOs at the top of each cluster directly to the objective. Any IOs that are not connected to any others, as well as the top IO in each cluster, are still needed to achieve the Objective, so they need to be connected directly to the Objective. These "hangers" are prerequisites for the objective and must therefore be tied below it. No IOs should be left without any connection after this step. All IOs should at least be connected to the Objective with their corresponding Obstacles.

Step 5—Check the Prerequisite Tree for Feasibility

This step ensures that you have sufficiently separated the objective into workable IOs to determine what actions we should take to achieve each Intermediate Objective. If you cannot think of the necessary actions to take, then additional obstacles must be present.

- Check the IOs that appear at the base of the tree—IOs with no arrows going into them. You should have actions in mind on how to achieve each of them. If you don't, it means that there are additional obstacles that you have not verbalized. If this is the case, ask yourself what are the obstacles that block me from achieving the desired IO? Select an IO for each, and connect them at the base of the tree to the IO in question.

- Read the tree both top-down and bottom-up.
  - **TOP-DOWN (starting with the Objective)** "I can't have (tip of the necessary condition arrow), until I have (base of the necessary condition arrow)."
  - **BOTTOM-UP (starting at the base of the tree)** "I must have (base of the necessary condition arrow) before I can get (tip of the necessary condition arrow)."
- Move into action. If appropriate, use a Transition Tree (TT) to develop your action plan, if you are using the PRT as part of a full TOC analysis. The Transition Tree should be completed to insure that your actions will lead to the achievement of the necessary IOs.

In the next section, we will discuss the basic principles of Transition Trees and how they can be used to develop action plans for difficult IOs.

TRANSITION TREES—BASIC PRINCIPLES

In the last section we discussed the Prerequisite Tree (PRT) with its many Intermediate Objectives. It is possible that you could encounter a particular Intermediate Objective (IO) that seems difficult to accomplish. When such an IO is encountered, the Transition Tree can provide the steps necessary to accomplish the IO. In essence, the TT can become the mini implementation plan for a specific IO on your way to accomplishing all of the IO listed in the PRT.

TTs are constructed using sufficiency-based logic and can be used to define and scrutinize the specific actions required to reach an Objective. When you are creating an action plan, most people focus primarily on the actions themselves, or on the desired outcome(s) of the action. Usually the focus is on "What are we going to do?" and "How are we going to do it?" rather than "Why are we taking this action in the first place?" It is important to remember that the primary focus of the TT is to focus your attention less on what you plan to do, and more on what you want to accomplish when determining and communicating the needed actions.

This is accomplished by coupling an action with a need to generate a desired effect.

This subtle, but important, shift in the focus, allows you to

- Monitor implementation progress by watching the effectiveness of the actions (meeting intermediate objectives along the way to the overall objective) versus just completion of the actions.
- Better make informed decisions and adjustments to the action plan, as required, instead of "going back to the drawing board" to re-write the entire plan.
- Communicate the key elements of an implementation plan effectively to others; the "What" and the "Why."

PRINCIPLES

The formal structure of a Transition Tree looks similar to a spinal cord, with the vertical stacking of Desired Outcomes. Because of this familiar "stacking," the core of the TT has been referred to as the "backbone." This backbone provides the description of the IOs that will gradually create the changes required. These changes will occur in reality, as a result of the planned actions. The TT methodology requires careful examination of the actions necessary to achieve the desired objective.

Parallel to the desired outcomes are the necessary actions. Each action is supported in sufficiency with additional causality of another desired outcome. When an ellipse supports an action and desired outcome, then sufficiency has been established. By scrutinizing each action, it can be determined if it is sufficient to produce the desired outcome and achievement of the TT Objective.

Often, we rely on a set of actions because "it worked for someone else," without checking if the actions really lead to the outcome you want, or if they fit your unique situation. When using a TT as a blueprint for implementing an objective, the focus is on causing specific changes in reality, rather than sticking to specific actions just because we think they will work.

Transition Trees can be used to

- Convert a strategic plan into a comprehensive tactical action plan.
- Plan important meetings, presentations, letters, phone calls, etc.

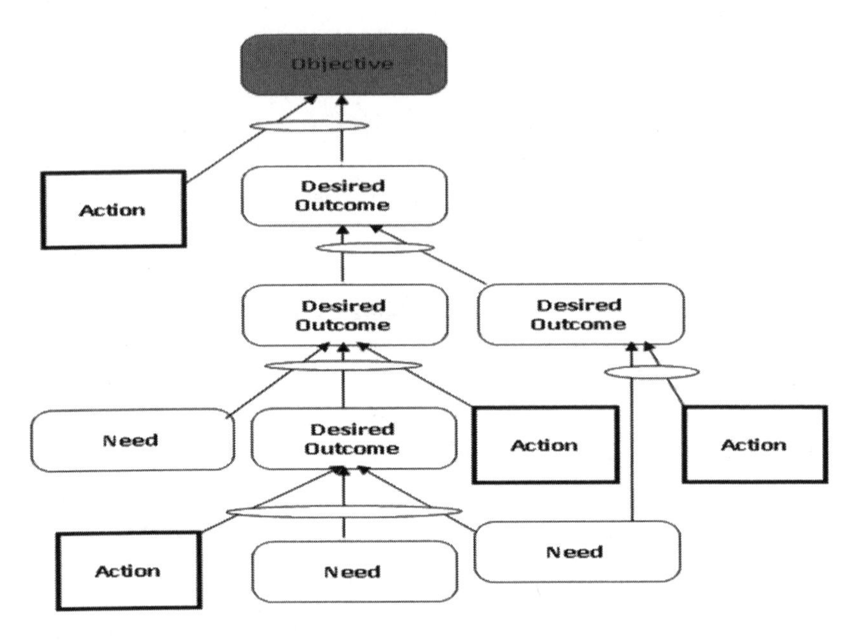

FIGURE 5.13
Basic Structure of a TT.

- Effectively delegate tasks through communicating both the "what" and the "why."
- Communicate how a sequence of desired actions will lead to specific effects.
- Solicit the needed collaboration of others.

The TT allows you to take your thinking to a very finite detail and determine the steps necessary to achieve the objective. The square "Action" boxes define the actions you must take, coupled with the needs you are trying to fill, to achieve the desired outcome you want. In the next section, we will discuss the steps for constructing a TT and developing the action plan to achieve the overall objective (Figure 5.13).

TRANSITION TREES—STEPS TO CONSTRUCT

In the last section, we discussed the basic principles behind a Transition Tree, what they can be used for and how they can help you achieve desired outcomes. In this section, we will discuss the steps necessary to construct the TT. There are five steps required to construct one, and if followed in sequence, they should yield positive results in accomplishing the desired outcomes to achieve the final objective.

Step 1—Define the Situation for Which You Are Preparing the Transition Tree

Not all situations can benefit from the construction and use of a TT. Before you invest the time and effort to build the tree, make sure it is something you really want to do. To benefit the most from a TT, you must choose a subject that is important to you and one in which it is helpful or necessary to map out a course of action. The TT will also provide a useful tool to define how the actions can lead to the successful accomplishment of the objective.

By clearly stating the starting point, you allow yourself the freedom to examine what it is you really want, instead of settling for what you think is possible. Write your situation in two sentences. First, describe the need, and second, define the subject of the TT you are going to create.

Step 2—Determine Your Objective(s)

A very clear verbalization of where you ultimately want your analysis to end up provides an effective means to evaluate the effectiveness of your actions. It is possible when building a TT to have more than one objective listed at the top. However, a TT with more than one objective is also harder to construct. It requires more focused attention on the Desired Outcomes for each objective. It is also possible that each objective could have similar or closely related Desired Outcomes. If the objectives are different, it could add confusion to your tree and make it difficult to explain to someone else. A word of caution is to be careful when selecting more than one objective. At times, it might also make sense to construct two trees, rather than just one.

When you are building a TT from a PRT, write down an Intermediate Objective from the PRT that appears particularly difficult to accomplish. It is possible to choose more than one IO from a PRT. However, it is more practical to develop a TT for each IO, rather than trying to combine them.

- *Ask yourself what the purpose of the plan is.*

  When the subject is a future event, the purpose is something that you want. When the subject is a past event, the purpose is the actual effects of the past. Sometimes a Transition Tree can have multiple purposes.

- *State the purpose as a specific objective in the present tense.*

  Most of the time, you will be writing a TT for attainment of a future objective. By verbalizing the objective in the present tense, you will be psychologically more invested in making it happen—it doesn't sound so unrealistic or difficult to make happen. This minor shift in perspective also helps you later in the process when you are checking the sufficiency of the logic. State the objective as clearly as you can.

- *Use the Transition Tree as a delegation tool.*

  There are many companies who proclaim "empowerment," yet there are few who have truly figured out how to make it happen. Delegation can be one way of enabling empowerment, as long as the person delegating doesn't override the decisions/actions of the delegates. The TT enables the one delegating to assign tasks and ensure that the delegates meet the objective with certain freedoms. The tree provides a succinct means of communicating clearly the "why" (the objectives/needs), along with the "what" (the task), so that the delegates understand more clearly the dynamics of the situation and are able to make decisions about actions, based on whether or not the objective will be achieved, without needing the constant support of the person delegating.

- *Hint: Look at the big picture.*

  Ask yourself how the objectives relate to the desired results of the organization. Will achievement of the objective help move the organization toward its goal? If it doesn't, you should probably reconsider what you have chosen as your objective and/or reconsider whether this is a good investment of your time.

Step 3—Determine the Actions Are Necessary to Achieve Your Objective(s)

Using cause and effect analysis, the future is always much more predictable than when using correlation techniques. You just need to determine what actions you need to take. In Step 3, you need to build the cause-effect structure that links the effects of your actions to the objective(s) you have selected.

When you think about the actions, you need to consider what things need to be done to make the objective happen. These actions define the grassroots level of things to be completed. These are the things that you can go do "right now." There is no planning below the action level. This is the place it starts.

You are seeking to logically bridge the gap between the current situation and the objective. The bridge will be provided through your use of cause and effect that provides the link between actions, their effects and the objective (Figure 5.14).

- *Write the objective from Step 2 at the top of a piece of paper.*
- *Write a connection that describes a step toward accomplishing the objective.*

The basic elements you are trying to connect include an action and an entity that describes the need for the action, and the desired outcome of the action.

Action: The action should be something that you believe is one of the first things required to reach the objective. State the action as a complete sentence as this will give you a starting point for constructing the tree. It may or may not be the first action at the bottom of the page when the tree is completed.

Need for action: There is something in the current situation you are intuitively aware of that has made you feel the need to take the action. After you verbalize the need, you have the opportunity to check if your

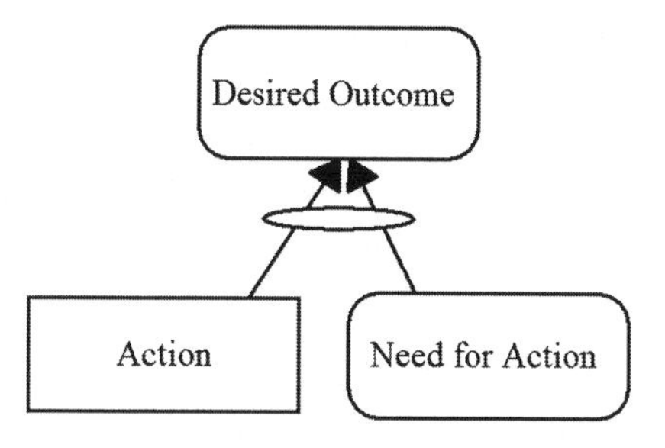

FIGURE 5.14
Link between Actions, Their Effects and the Objective.

assumptions about reality are valid (i.e. apply the Entity Existence reservation). This also provides an anchor by which others can judge the need for the action.

Desired outcome: What do you need to have to achieve the objective? Usually people fill this in after they have verbalized the action, so the question becomes, What do you need to gain from the action, so that you are closer to achieving the objective? This step is difficult for many because we all know that there are many possible effects that may result from an action. What we are trying to do through the Transition Tree is have some control over the possible effects, by causing just one to occur. Clearly state what it is that you want the action to accomplish.

- *Solidify the causality.*

  This step requires that you carefully check your logic using the Categories of Legitimate Reservation and fill in any holes you can find. Add actions as needed to ensure the desirable outcomes. This is the step at which you must be very precise about what you have written. An analysis on paper may look good to you, but reality is what counts.

- *Continue building upward until you have reached your objective (stated in Step 1).*

  At this point, you probably have some "blank" space between a Desired Outcome and your objective. Starting from the Desired Outcome (Step 3), ask yourself what needs to be done next that will bring you closer to the objective. If you are an action-oriented person, it will probably be easier for you to construct the tree from the action perspective. If you think more in terms of goals, it will probably be easier for you to continue building from the Desirable Outcome perspective. Regardless of the trigger source, you will need to continue to verbalize actions, needs, desirable outcomes, that help solidify causality.

Hint: Sometimes you will choose as an action something you don't know how to do, or something that someone else is required to do. These are not your actions but "desired outcomes," which still require some action on your part to achieve. Change such desired outcomes to round boxes, and beneath them add the action you need to take to achieve it.

Some people might experience difficulty continuing to build the tree (Figure 5.15). If such is the case, then use the following template to help fill in the missing links. This template can be especially helpful to review

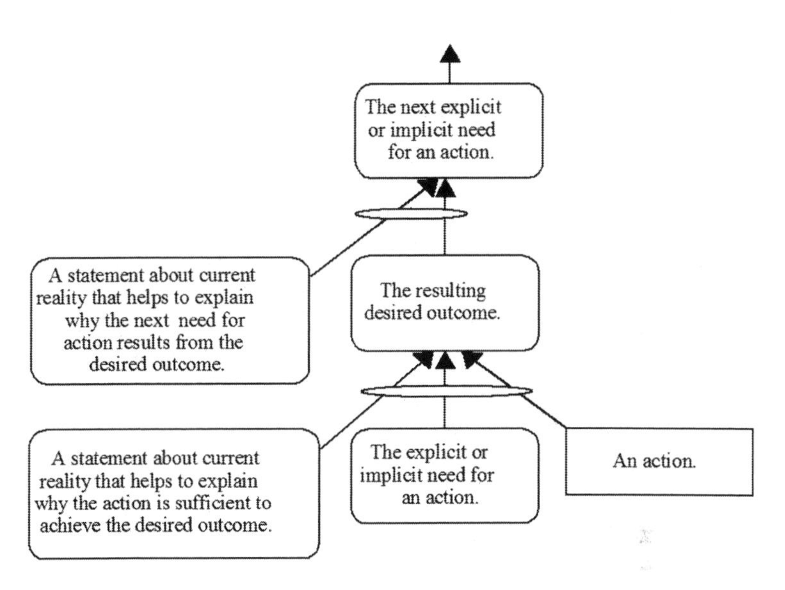

FIGURE 5.15
Template to Help Fill in the Missing Links in TT.

when you are seeking to modify the structure for those who will read it, as it forces the writer to verbalize some steps in the evolution of the tree that they may have taken for granted.

Step 4—Look for Any Undesirable Side Effects

For most people, it is very easy to convince themselves that their plan will work. However, many plans have been known to fail, especially when some element of your plans does not produce the desired effect, and in fact produces an Undesirable Side Effect. Undesirable effects can sometimes happen quickly and have instantly devastating effects on your planning. Doing this step will guide you through the process to identify the potential pitfalls of the plan and help define the necessary actions then to resolve them.

- *Ensure there are no undesirable side effects.*

  For most people, it is very easy to convince themselves that their plan will work. However, many plans have been known to fail, especially when some element of your plans does not produce the desired effect and in fact produces an Undesirable Side Effect. Undesirable effects can sometimes happen quickly and have instantly devastating effects on your planning. The steps of Key Action—4 will guide you through the process to identify the potential pitfalls of the plan and help define actions then to resolve them.

- *Examine possible negative effects of your actions.*

  Start at the action appearing at the bottom of the page and ask yourself "If (action) and everything else remained as it is today, then ...." This will surface potential effects of the action, some of which are minor and others of which might be major. If you find something major, proceed to the next section. If you don't find anything, examine the next action that appears in the tree, keeping in mind that you have created a partially new reality. This will slightly impact the question you will ask to surface additional effects: "If (action) and (new reality) and everything else remained as it is today, then...."

- *Add actions that will eliminate undesired effects.*

  Focus your effort on removing the major undesirable effects that you have surfaced. Figure out what action can be taken to eliminate it. Incorporate this action into the tree. The only way to eliminate an unwanted effect is to take action—just saying that it will not happen will not make it so. Don't forget to check if your new action causes other major negative effects.

You are finished with this process once you are satisfied that you know the potential side effects and you have resolved all major issues that could jeopardize it.

Hint: Here are some questions to ask that may help surface additional negative branches:

- What is the impact of this on T, I and OE?
- What is the impact of this on other parts of the organization?

Step 5—Take Action

Reality doesn't change with thoughts and plans, but rather with your actions. Step 5 is a gentle reminder that for the objective(s) to be realized, actions must be taken in reality.

There you have it. I hope these writings have been somewhat sufficient to give you a better understanding of the TOC Thinking Processes and the ability to apply some of what you have learned to problem solving.

REFERENCE

1. Dettmer, H. William, *The Logical Thinking Process—A Systems Approach to Complex Problem Solving*, American Society for Quality Control Press, Milwaukee, WI, 2007.

6

A Simplified Improvement Strategy

INTRODUCTION

When Eliyahu Goldratt [1] first exposed the world to his then radical Systems Thinking Processes, he presented them as the logical tools to express, and document, logical thinking in a very structured format. The improvement quest was centered on the ability to answer three questions:

1. What do I change? (What problem are you trying to solve?)
2. What do I change to? (What is the best solution to solve the problem?)
3. How do I cause the change to happen? (Implementing the solution into reality.)

Goldratt developed the five Systems Thinking tools to accomplish the necessary thinking tasks to succinctly solve a problem. Each of the five thinking tools aids the user with the ability to apply the tool as a singular component, or to use the tools in a sequential combination, to strengthen the logical analysis even more. The five Systems Thinking tools are

1. Current Reality Tree (CRT)
2. Evaporating Cloud (EC)
3. Future Reality Tree (FRT)
4. Perquisite Tree (PRT)
5. Transition Tree (TT)

Today, these five tools still remain as the foundational cornerstone of the Thinking Processes for the Theory of Constraints (TOC). These thinking tools have, through time, proven their worth to better understand and analyze simple as well as complex problems. For any of you who have been through a Systems Thinking Processes course and learned how to use these tools properly and effectively, you understand "how" and "why" a good analysis can command a rigorous and sometimes sustained effort to accomplish the task. In other words, developing a useful and solid analysis using the Thinking Process Tools can take some considerable effort. In some respects, the time commitment required to do a good systems analysis has been a downside to using the thinking tools and has, in turn, caused many people to ignore the tools and turn their heads to the real power and usefulness they can provide.

Over the years, global dynamics have evolved to such a level of forcefulness that instant gratification is now considered the norm. That is, put your nickel in and get something out now! The dynamics of this global phenomenon have pushed us all to chant the mantra of "Better, Faster and Cheaper!" Waiting any amount of time for something to happen no longer seems to be an acceptable option—it's the world we live in. There are situations and problems that are not accompanied by the luxury of the required time to figure them out. This requirement to be "Better, Faster and Cheaper" has not gone unnoticed in the network of the world's systems thinkers.

THE EVOLUTION OF THE THINKING PROCESSES

If an idea is presented to the world, and it's a good idea, then somebody, somewhere, somehow, will improve the idea and expand it to another level. The "improved" idea will overcome some of the prior inertia of the obstacles and issues that seemed to exist. If the idea is really good and accepted by many people, then the evolution of the idea continues. Each new level of improvement removes more of the obstacles from the previous level, and each new level is presented as a form of unification of ideas from the previous level(s). This unification notion becomes the idea of doing more with less, or combining for better results. For instance, instead of doing three separate tasks, now you do one task and get better and faster results. In other words, through the steps of unification, an idea can now become easier and more usable by more people. Such is the case for the TOC Systems Thinking Processes—a good idea that continues to evolve.

THE APPARENT PROBLEM

As a person who has taught many Systems Thinking courses (i.e. Bruce Nelson), I've had the opportunity to present the Systems Thinking tools to a wide variety of people. The intellectual levels, the passion and the job functions have been spread over a variety of individuals and industries. Teaching these courses has also delivered a wide range of results. In some classes, there was 100 percent completion, while in other classes results were dismal with a 75 percent to 80 percent failure rate. Failure rate in this case is defined as those students who did not finish the course, which was usually related to the time commitment required.

In the early days of teaching this course, the preferred approach was to provide a level playing field for all students and make it as linear as possible. The desire for doing this was to reduce as much variation as possible in the learning environment. However, even after creating the utopian venue, there were some surprises. In a typical class the students would be divided into teams, so that each team had a minimum of two members and in some cases three or more, if required. Each team was given a two-page write-up (case study) about a fictitious company that was having problems. The assumed linear thinking was that everyone who read the paper would discover the same problems, and ALL readers would eventually reach the same conclusions for the Undesirable Effects (UDEs). Such was not the case. In a class with ten students and five teams, it was very predictable that these five teams would develop five different core problems when they constructed their Current Reality Tree. It is a ponderous thought to speculate how it was possible for five different teams, each analyzing the exact same problem, to come up with five different answers. It was also a surprise to discover that in most cases, each of the five different answers could be plausible! There was also another observation—the confidence level of the students in thinking they had correctly discovered the core problem was absent. It seemed that the constant question from the group was, "Is this the right core problem?" Even when a core problem was stated (defined), the lingering question became, "Is this REALLY the core problem?"

When you apply a truly scientific method toward problem solving, then a potpourri of answers should not be possible. In fact, only one answer should be correct! And yet, at the same time, it appeared as if more than one answer could, in fact, be correct. Why? The answers provided by the students were not the same, and yet they all seemed to be related. It seemed that what was actually being exposed was a listing of the "obstacles" or "interferences" that stopped them from achieving what they wanted! All of the answers (the perceived core problems) presented were plausible reasons (interferences) to better understand "why" what they wanted more of could not be achieved. This epiphany created a path back to a thinking tool that we had previously used, the Interference Diagram (ID).

THE INTERFERENCE DIAGRAM

When the Interference Diagram was first being developed and drawn on whiteboards, it was done so not with the intention of replacing or linking any of the current Systems Thinking tools, but rather to fill the void of a necessity, for completing the analyses in less time. The ID is a thoughtful mind mapping tool that can quickly point a team, or individual, in essentially the right direction to solve a problem and not be required to construct a Current Reality Tree. In essence, the Interference Diagram was able to answer the question "What to change." In this case, the "What to change" became a list of the many entities (Obstacles/Interferences) and not just a single entity (Core Problem).

The first uses of the ID came from Bob Fox [2] at the TOC Center, New Haven, Connecticut, circa 1995. Since then, the use and structure of the ID tool has not been well documented, published or transferred to the public at large, but rather used by a limited number of practitioners within the TOC network. The simplicity of the ID and the underlying robust concept to solve problems has been applauded by ID users. The global influence of the ID to solve problems is colossal. Unlike the other thinking tools, the ID is not based on logic, but rather on intuition. The arrows in this diagram are just arrows without sufficiency or necessity. The applied thinking is not to develop or isolate a single answer, but rather to list the obstacles/interferences that block the achievement of the desired objective.

It has often been said that the biggest obstacle to solving a problem is to first be able to precisely define the problem. If you are not sure what problem you are trying to solve, then it is awfully difficult to determine the correct solution. In other words, if you don't know where you are going, then any path you decide to take is sufficient to get you there—you'll just never know when you get there. The ID structure and concepts are very simple, and yet very powerful in the results provided.

INTERFERENCE DIAGRAM TYPES

There are actually two different ways to apply an ID. The first application is using the ID as a thinking tool to exploit a known constraint. The second application involves using the ID in combination with the Intermediate Objectives (IO) Map (aka Goal Tree). The second application offers a fast and highly effective way to develop an overall strategy plan and implementation plan. We will discuss the second application later.

First, let's consider the exploitation of a constraint within a system. When the system's constraint is identified, then the exploitation question becomes, "How do I get more from the constraint?" What are the "interferences" that slow down or stop the constraint from doing more and or doing better? It is possible that there could be several interferences that block the enriched performance of the constraint. This "interferences" listed become the reasons "why" the constraint cannot do more. The interference list is best compiled from the resources that use the constraints. The constraint users can provide the subject matter expertise to define the interferences and are most familiar with the constraint and how it works, or doesn't work. When constraint users are asked the question, "What stops (interferes with) you getting more from this operation?" Chances are good that the user resources will be brutally honest with their answers. What becomes important at this stage of information gathering is to filter the "emotional" response from the "logical" responses. You'll need to determine if the response is really a system problem or strictly a personal annoyance or gripe. This analysis will provide better results if the emotional responses are removed up-front, before placing the statement on the interference list.

The entities on the list simply imply those "interferences" that stop the constraint from doing more. That is, those interferences that "steal" time away from the constraint. To get more from the constraint, you must reduce the impact of the interferences, or remove them completely. Any interference that can be reduced or eliminated, will free up additional time for the constraint to work more. As an example, let's go through the building steps to construct an ID that is used to exploit a constraint in a production system. In our example, let's say that we have identified one particular machine in a production line as the constraint. We will refer to it as "Machine XYZ." Let's define the steps to construct.

Step 1—Define the Goal/Objective

The goal/objective should be something that you really want, but something that doesn't exist in your current reality. For our example, we will choose as the objective "More parts from the XYZ machine." This can be written on a whiteboard or in the center of a piece of paper.

Step 2—Define the Interferences

Step 2 is best accomplished using observation of the system and interviews with the operators. When observing the constraint, look for those things that slow down the constraint or stop it from working. What are the interferences that take time away from achieving more of the objective? If the identified constraint is truly a system constraint, then keeping it busy all the time and getting more output will be paramount to successfully gaining more system throughput. A possible list of obstacles/interferences might include

1. Parts not available to work.
2. Operator on break/lunch.
3. Operator has to find his own parts.
4. Operator is looking for the Supervisor.
5. Operator is attending training.
6. Machine is broken.

The list could be extensive and varied, but what is most important is to identify those things that stop the constraint from doing more. Observation of the constraint might reveal other things that impact time at the machine, such as having to do setups for a different product. All of the observation and interview items combined equal the list of obstacles/interferences that hinder achieving the objective "More parts from the XYZ machine." There is not a set limit to the number of interferences that need to be gathered, but rather, you should list as many as you think necessary to fully describe "why" the machine stops working.

Step 3—Quantify the Time Component for ALL Interferences

Quantifying the time component associated with the interferences becomes important to fully understand and appreciate the impact of the interference on the available time. The time component will help filter the important few from the trivial many and help express those items with the greatest impact. Knowing the impact of the time component will also be useful in determining the priority ranking for which interferences to reduce, or eliminate, first. Some of the interferences will be more important than others, because they are not all equal. When you accurately quantify the time component for the interference, it also allows for excellent Pareto analysis. Pareto analysis will align the interferences based on time distribution and determine which interference is most impactful, and the intrinsic order of improvement. Pareto allows the focus necessary to gain the most leverage from the action implemented. However, it is also not realistic to assume that all interferences can be reduced and/or eliminated. There will be some interferences that do not offer themselves as candidates for elimination but rather as entities that can benefit from a reduced time impact. In other words, if an entity with a time impact of 45 minutes can subsequently be reduced to 15 minutes, then the benefit gained for the system is 30 minutes more time for the constraint. In other cases, the interferences cannot be reduced or eliminated at all. In our example, the time for breaks and lunch cannot be removed because employees are, by law, allowed lunch and break time. However, as an alternative you could gain some machine time by having an alternate person or crew work the machine during lunch and breaks.

What you are really looking for in Step 3 is to quantify, with time, those activities that are stealing time away from your constraint operation. If you can eliminate, reduce or off-load some of these activities, then more time is available to get more parts through the constraint. If interferences are known, and corrected, then the end result should equate to more output from the machine and more throughput through the system.

Step 4—Alternatives to the Interferences

The interference list defines all of the obstacles/interferences that stand in the way of having more of what you want. If these interferences/obstacles did not exist, then achieving the goal would be easier. With the interferences defined, you should be able to counter the seemingly negative effect of the obstacles/interferences with an Injection/Intermediate Objective. This action can be accomplished by asking the question, "What must exist, so that the interference no longer exists?" Whatever your answer might be is the Injection/Intermediate Objective to overcome the negative impact of the interference. Continue working your way down the list, and create an Injection/Intermediate Objective for each Inference/Obstacle listed.

The items on the Injection/Intermediate Objectives list are the things that must be accomplished to reduce or eliminate the negative effects of the inferences. If not, then consider revising your Injection/Intermediate Objectives list until the answers are sufficient to remove the obstacles/interferences. With the addition of Injections/Intermediate Objectives, the list should provide sufficient "ideas" to move you closer to the objective/goal you have established. With the list in place, you have now precisely defined "What to change to."

Obstacles/Interferences Intermediate Objectives/Injections

1. Parts not available to work.	1. Parts are kitted and ready for use.
2. Operator in on break/lunch.	2. Train an alternate crew or person.
3. Operator has to find his own parts.	3. Parts delivered to operator.
4. Operator is looking for the Supervisor.	4. Supervisor notification system.
5. Operator is looking for paperwork.	5. Paperwork follows job through the system.
6. Machine is broken.	6. Preventive maintenance (priority #1).

Figure 6.1 presents an example of what a completed Interference Diagram might look like for our example. The center circle contains the objective, and each of the interferences are listed around a circle. The direction of the arrows makes no difference, because these arrows are based on the ID user's intuition and not necessity or sufficiency logic.

Notice in Figure 6.2 that the interference times have been added to each interference identified. This allows for a quick visual of the time impact.

When using the Pareto analysis, you should base it on available machine time, because this will succinctly show the impact of the interferences. If the available machine time for the XYZ machine is considered to be eight hours, for one shift, then the available time is 480 minutes (60 min × 8 hours = 480 minutes) during an eight-hour period. When you use the 480

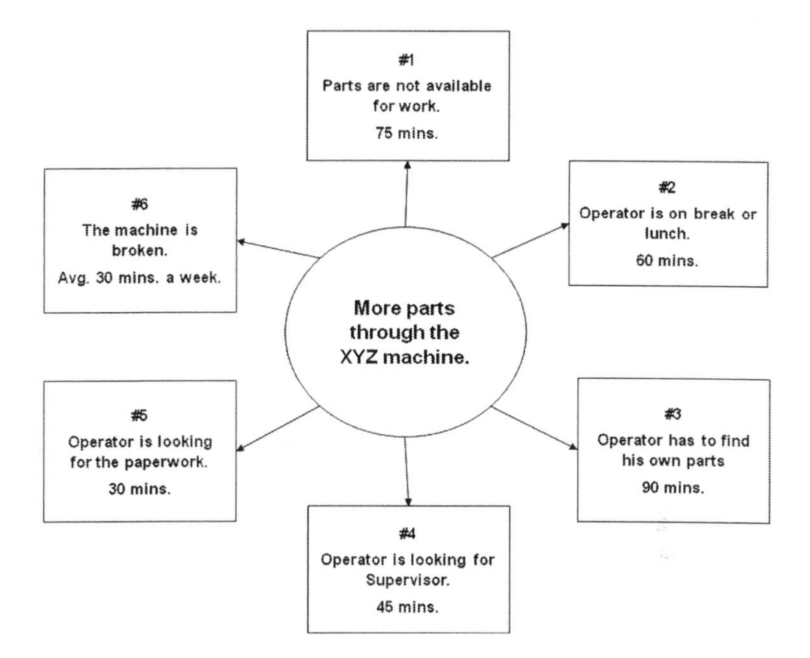

FIGURE 6.1
Example of What a Completed Interference Diagram Might Look Like.

#	Description	Daily minutes	Percent Interference
1	Parts not available	75	16%
2	Breaks & Lunch	60	13%
3	Operator finding parts	90	19%
4	Looking for Supervisor	45	9%
5	Looking for paperwork	30	6%
6	Machine is broken [1]	6	1%

Total Interference minutes [2]	306	
Available work minutes [3]	174	
Total Available minutes	480	
Utilization Percentage		36%

[1] 6 minutes is calculated from a weekly average of 30 mins (30 mins/ 5 days)

[2] $\sum$ interference minutes

[3] Available mins - Interference mins.

FIGURE 6.2
Interference Times Added to Each Interference Identified.

minutes as the baseline to measure the impact of the interferences, the results will show the conflict between the time available to work, and the time the machine actually spends working. Figure 6.2 shows an example setup of the Excel input sheet, to show the interference description, total minutes the interference consumes, and the percentage impact on the total minutes available. As you can see the leading time impact comes from not having parts available at the machine to work on. Fixing this interference alone could provide an additional 90 minutes of throughput time every day.

From this spreadsheet setup, you can display two additional charts that help drive the point home for the impact of the interferences. With the interference percentage calculated, you can create a pie chart to visually display the breakout and "Interference Impact." Figure 6.3 displays the Interference Impact for our example.

Figure 6.3 shows that only 36 percent of the total time available is actually being used to make parts on this machine. The other 64 percent of the time is consumed by the interferences. Figure 6.4 displays the interferences in a Pareto analysis, to show the intrinsic (descending) order of improvement. Those interferences listed highest on the list should be reduced or eliminated first, if possible, to gain the most benefit.

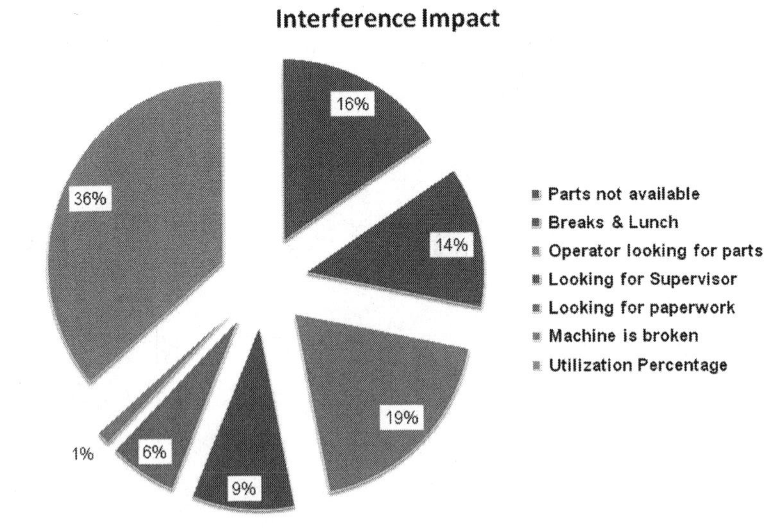

FIGURE 6.3
The Interference Impact.

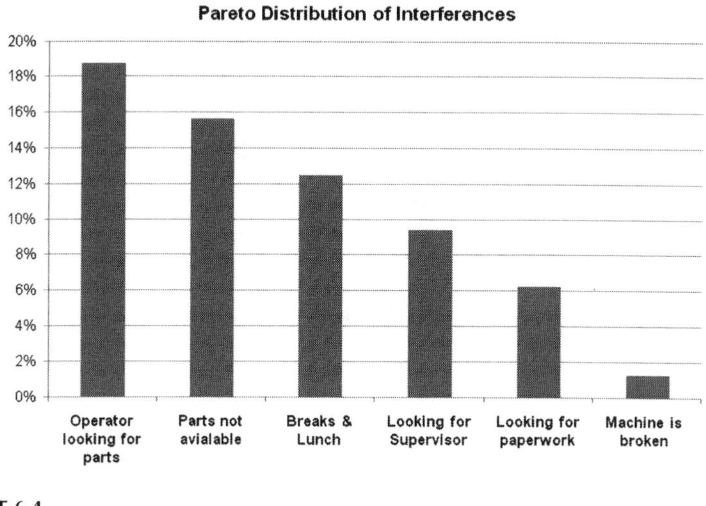

FIGURE 6.4
Pareto Chart.

INTERFERENCE DIAGRAM FOR STRATEGY

In the previous example, we demonstrated how to use the ID to exploit a constraint, so now let's look at the second application, strategy development. When using the ID for strategy development, you want to focus on a higher level (more global) Objective/Goal, especially if the input audience is cross-functional. The Objective/Goal has to be at a high enough level to satisfy all parties concerned. If you drive the Objective/Goal too low, some organizations will complain that it does not apply to them. When using the ID for the purpose of strategy development, it is best if used in tandem with the Intermediate Objectives Map. The combined thinking power of the ID/IO Map is referred to as the *Simplified Strategy* and will enable an effective strategy, and detailed implementation plan, to be developed. In this case, the IO Map provides the intrinsic order (sequence) of the tactical actions to accomplish the overall strategy.

When the ID is used in a strategy development scenario, it is not a constraint you are looking for. Instead, what you want to define is the strategic Objective/Goal that you really want to achieve. When the Objective/Goal has been defined, then look next for the "interferences" that slow down or stop the progression toward the goal. Usually, there can be several interferences that block the path to success. If there is one thing people are really good at, it's their ability to express their vocal rationale about "why" something won't work, or why you cannot have something you want. When you present your idea to someone else, or to a group, the common reaction is, "that's a great idea, but ...." As soon as the person says "but," they will interject the reason "why" they think the idea won't work. "It's a great idea, but the boss will never approve it." Or, "That's a good idea, but it's not the way we do it here." Or maybe even, "It's a good idea, but it will be too expensive to do." What they are telling you is what they think the interferences or obstacles are, that will hinder your ability to succeed. To get more of what you want, you must reduce the impact of the interferences or obstacles, or remove them completely. As an example, let's go through the steps to construct an ID that will be used for strategy development.

Step 1—ID Strategy—Define the Goal/Objective

The Objective/Goal for the strategy application of the ID is focused on strategic direction, rather than a constraint, and generally answers the question "Where do we go from here?" The objective has to be at a level high enough to include the many. However, with that said, it is also important to understand that good strategy development can happen within a specific organization, such as Engineering, Procurement or Manufacturing. Think in terms of a higher-level Objective/Goal for this scenario when dealing with a single organization, or combining the objectives of many organizations into a single strategy at a higher level. The Objective/Goal that you pick might seem elusive, but it should also be necessary to get where you want to go, the focused end point of the journey.

The ID tool can work well in a group setting and allow the user to surface obstacles and interferences across many different organizations. Suppose for our example we defined the objective as "Increasing Revenue."

Step 2—Define the Obstacles/Interferences

When a group consensus is realized, and the objective has been clearly defined, then you need to look for the obstacles or interferences to characterize "why" you can't have what you want. Again, when using the ID for strategy development, the obstacles or interferences can, and most probably will, cross many organizational functions. There is no minimum or maximum number of obstacles or interferences required. Rather, you should be looking for a list of interferences that is comprehensive enough to surface those entities that are really standing in the way of your success. Suppose, for this example, our cross-functional team listed the following obstacles and interferences for achieving the Objective/Goal of "Increased Revenue." The obstacles/interference list might look something like this:

1. Not enough sales
2. No markets to grow into
3. Customer has low perception of our product
4. Products are priced too high
5. Competitor has higher quality
6. Production takes too long

This list defines the obstacles and interferences that are perceived to currently exist that block successful achievement of the goal. These are the things that stand in the way of being able to achieve "Increased Revenue." This list can, and probably will, seem a bit overwhelming when you look at it, but don't lose faith just yet. Let's do the rest of the steps and see if we can tame the beast.

Step 3—Define the Intermediate Objectives/Injections

As part of the ongoing group discussion, you'll want to define the Intermediate Objectives/Injections. In Step 3, what you want to surface are those IOs that must exist to make the obstacles and interference go away and not be a problem anymore. Ask the group this question, "What must exist in order for the obstacle or interference not to be a problem anymore?" When you think of the actions required for eliminating or reducing the obstacles or interferences, be bold. Describe what you think is really necessary, to counter the problem(s) to accomplish the stated objective. Don't shy away from Intermediate Objectives/Injections, just because you think you cannot make them happen. If they are important for the overall objective, then write them down, no matter how far out there they might seem. Here are some possible intermediate objective examples for the obstacles on the list:

Obstacles/Interferences	Intermediate Objectives/Injections
1. Not enough sales	1. Increased sales
2. No markets to grow into	2. Explore/find new markets
3. Customer has low perception of our product	3. Customers have high product perception
4. Products are priced too high	4. Products priced competitively
5. Competitor has higher quality	5. Quality higher than competitor's
6. Production takes too long (lead time)	6. Lead time reduced (increase throughput)

With the Intermediate Objectives/Injections defined, you now have the list of the required actions to eliminate the obstacles/interferences and make them go away. It is a forbidding list, and at first glance, it appears almost impossible to achieve any of these actions. Figure 6.5 provides an example of the basic structure and layout for using the ID for strategy development.

With this section complete, the next step is to lay out and construct the IO Map, to govern the implementation plan and determine the tactical actions to accomplish the objective, which we will cover later.

When the ID is used to analyze and exploit a constraint, it can be a quick and effective tool to generate good ideas quickly. This method, when used in conjunction with Pareto analysis, can quickly provide the visual tool to determine the impactful interferences and provide the focus and

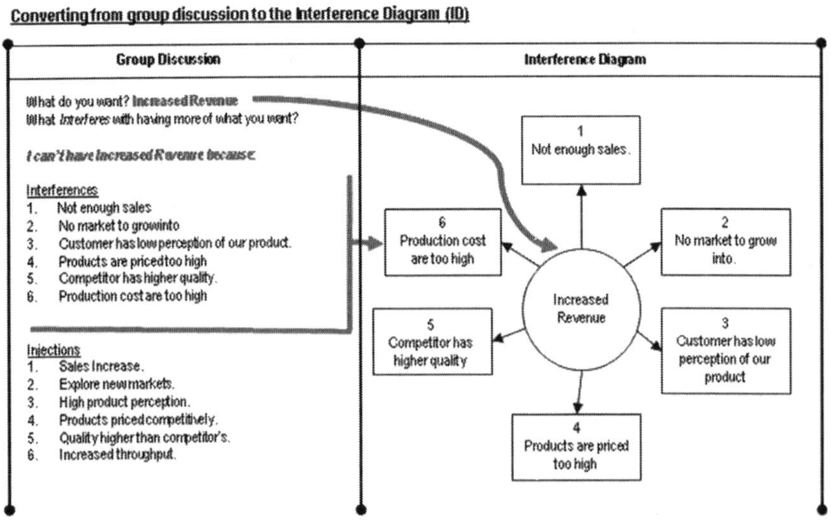

FIGURE 6.5
Converting from Discussion to an Interference Diagram (ID).

leverage on those few important actions that will provide the highest levels of improvement.

INTERMEDIATE OBJECTIVES MAP

The Intermediate Objectives Map was developed by H. William (Bill) Dettmer [3], and in the spirit of combining methods (Unification) within a methodology, this tool fits the criteria. The IO Map has been refined over the years to become a very practical and useful organizational and thinking tool. Instead of using the full spectrum of the Thinking Process tools to conduct an analysis, the IO Map combines the Prerequisite Tree (PRT) and the Conflict Diagram (CD) into a single tool. In his paper about Intermediate Objective mapping, Dettmer defined the IO Map as a Prerequisite Tree (PRT) without any obstacles defined.

Dettmer's primary intent for this tool was to simplify the construction and accuracy of the CRT, to focus the attention on a better defined objective, rather than a core problem. In this context, the IO Map can be used to surface the Undesirable Effects. In most cases, the UDEs can be discovered by verbalizing the exact opposite of the desired intermediate objectives that are listed. These UDEs then become the building blocks for a Current Reality Tree. It makes sense that UDEs can come from the IOs. UDEs are what currently exist, and the IOs are what you want to exist. In essence, the IOs are what "you want," and the opposite wording of the IOs would embrace the UDE list.

Dettmer also defines an expanded function for the IO Map, indicating that it serves well the purpose of strategy development, and as such, it does provide a robust tool to do that. The IO Map, used as a stand-alone technique, can provide the necessary clarity and direction to accomplish a needed strategy.

The IO Map is a very concise organizational thinking tool based on *necessity logic*. The IO Map allows the user to define the IOs, and then the intrinsic order of the IO task completion, by using necessity logic. In other words, it is read with necessity as the outcome. *In order to have … (Entity Statement at the tip of the arrow), I must have … (Entity Statement at the base of the arrow)*. Necessity logic states, in essence, that entity B must exist before you can have entity A. The entity cannot be there just sometimes, or most of the time, but instead necessity states it *MUST* be there. The existence of entity B is not a causal existence. Necessity requires that the "B entity" exist before the "A entity" can be achieved.

The structure of the IO Map is really very simple. There are three primary levels, or thinking levels, required to construct the IO Map. The first level is defining the *GOAL*. The second level identifies the *Critical Success Factors* (CSFs), or those intermediate objectives that must exist, prior to achieving the Goal. The third level is populated with the remaining *necessary conditions* (NCs) required to achieve the Critical Success Factors. Figure 6.6 provides an example of the basic structure of an IO Map.

Figure 6.7 shows an example of the IO Map that was created using our IO list example.

You'll notice the shaded IOs in the diagram. These are the IOs that were surfaced when building the IO Map. These IOs did not appear on the original IO list, but instead surfaced after construction began on the IO Map.

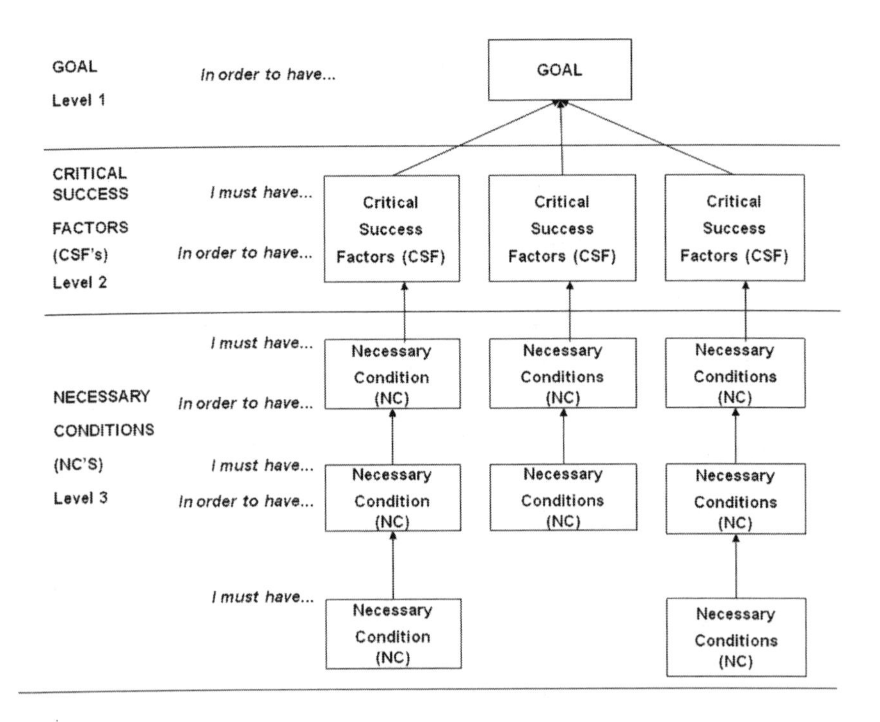

FIGURE 6.6
Completed Goal Tree/IO Map.

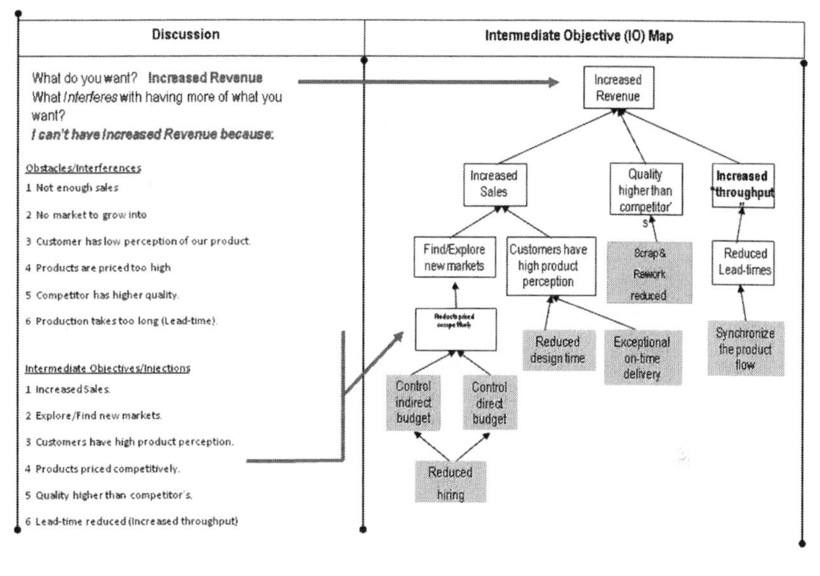

FIGURE 6.7
Example of the IO Map That Was Created Using Our IO List Example.

THE ID/IO SIMPLIFIED STRATEGY

Now that you have an understanding of both the Interference Diagram and the Intermediate Objective Map, and how they can be used as stand-alone techniques to generate some impressive improvement results, let's talk about how they can be combined.

The Simplified Strategy is a way to combine these two tools, depending on the situation being analyzed and the desired outcome required. It is possible that when using the ID to define the interferences, they are actually obstacles that are not necessarily time driven, but rather event driven. The ID allows you to define the obstacle/interference if they are not already well known. Sometimes, the obstacles do not provide the means to implement a simple solution in isolation, but rather are collectively connected by necessity. In other words, when you develop the list of obstacles using the ID, the IO list becomes the verbalization opposite of the obstacle rather than just an injection. You are looking for the IOs that must exist in reality to make the obstacle/interference not a problem anymore.

What happens next is the listing of IOs becomes just that, a list of IOs. Now, with the IO mapping tool, you can establish the logical necessity between single IO events (entities) that requires another predecessor event (entity,) before the event can happen. In other words, there is a logical dependency and intrinsic order in the sequence. Just randomly selecting and completing IOs will not satisfactorily achieve the goal. When you analyze the IO list you realize that ALL of the IOs need to be completed, but which one do you start with? When this is the case, the IO Map can be used to determine the sequence and order of completion.

From the IO list, you can determine which events are Critical Success Factors and which ones are the NCs.) By using the IO Map to determine the necessity between the events, it becomes *exactly* clear which IO you need to start with to implement your strategy. Each level of the IO becomes logically connected, to formally outline the "strategy" and "tactics." In other words, the goal is the strategy, and the Critical Success Factors are the tactics to accomplish the goal. At the next level, the Critical Success Factors become the strategy, and the necessary conditions become the tactics. The same thinking applies down through the next levels of necessary conditions. When you reach the bottom of an IO chain, then you know what action you need to take first, to start the process moving up through the IOs. By using the IO Map as a problem-solving supplement to the ID, it provides the needed organization to logically align the IOs. The ID map will provide the well-defined obstacles/interferences, to better focus the creation of the correct IOs to negate the obstacles. Consider also, that sometimes it is very difficult to generate a good solution without first understanding what exactly the problem is.

Even though these tools can be used in combination, it might not always be necessary to do so. In fact, the power of these tools allows them to be used in reverse order if so desired. If you already understand what you need to do, then the IO Map can be your beginning tool. If, however, you are not so sure why you cannot achieve a particular goal, then the ID helps identify the obstacles/interferences. Even if you begin with the IO Map and you discover a particular IO that is necessary, but you're just not sure how to make it happen, then you can use the ID as a subset of the IO Map to discover the interferences for achieving that IO. If you remove the interferences for the IO, then you can achieve the IO. When you achieve that particular IO, you can move on to the next one. If you already know how to accomplish that IO, then fine. If not, then use the ID again to surface the interferences.

Figure 6.8 shows a possible template for the combined approach.

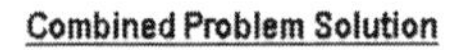

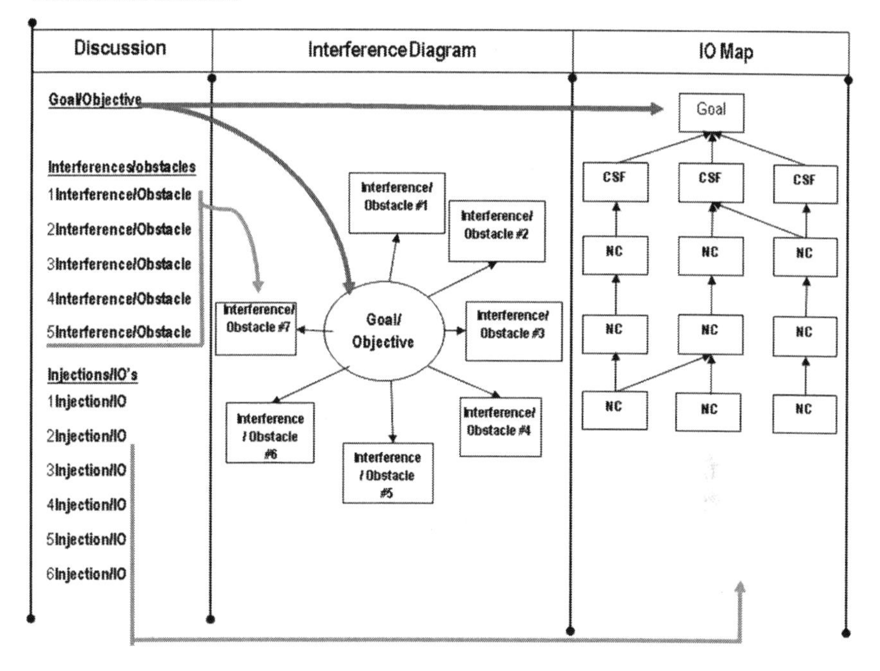

FIGURE 6.8
Possible Template for the Combined Approach.

With the ID/IO Simplified Strategy complete, you now have the outline necessary to prepare an effective and accurate Implementation Plan [4]. The intent of the IO Map is not to provide implementation detail at a low level, but rather to provide milestones or markers to make sure you are walking the right path. For each IO listed you can provide the required detail about how it will be completed.

CONCLUSIONS

In a world that requires "Better, Faster, Cheaper," the Simplified Strategy approach of the Interference Diagram and Intermediate Objective Map can provide exceptional results in a shorter period of time. By combining the power of these thinking tools, the user will benefit from an effective and complete analysis that is completed in significantly reduced time. These tools, used in either a stand-alone environment or a combined approach, will provide the thinking necessary to develop good results. The speed with which these tools can be used is an enormous advantage over the original Systems Thinking tools, allowing the ability to answer the three questions:

- What do I change?
- What do I change to?
- How do I cause the change to happen?

The structure and concept behind these tools makes them easily adaptable and well understood and accepted in a group situation, to allow for faster collection of data and analysis of issues.

REFERENCES

1. Goldratt, Eliyahu and Cox, Jeff, *The Goal—A Process for Ongoing Improvement,* North River Press, Barrington, MA, 1992.
2. Fox, Robert, TOC Center, New Haven, CT, discussions, circa 1995.
3. Dettmer, H. Wlliam, "The Intermediate Objectives Map," http://goalsys.com/books/documents/IOMapPaper.pdf, November 2008.
4. Nelson, Bruce H., "ID/IO Simplified Strategy" (CS^2), Original works, March 2011 Copyright © 2011 by Bruce H. Nelson. All rights reserved.

7

Project Management

INTRODUCTION

In a fairly recent survey (The Chaos Report) by the Standish Group, studying nearly 10,000 Information Technology projects across America, it was reported that 52 percent of projects ended up costing greater than 189 percent of the original budget, 31 percent were cancelled and only 16 percent of the projects were completed on time and on budget.

In this chapter, we are going to write about project management. Specifically, we're going to compare the primary method being used today, called the Critical Path Method (CPM), to a Theory of Constraints (TOC)-based method known as Critical Chain Project Management (CCPM). Before we make this comparison though, we want to write briefly about the state of project management in the world today.

In 2007, a new Chaos Report revealed that only 35 percent of software projects started in 2006 can be categorized as successful, meaning that they were completed on time and on budget and met the user requirements. Although this was a marked improvement from their first groundbreaking report, it's safe to say that these statistics still aren't acceptable or at least, where they need to be!

Yet another study, done in Australia, found that construction projects completed only one-eighth of building contracts within the scheduled completion date and that the average overrun exceeded 40 percent. The fact is, there are many other reports from numerous industry types, from all over the world, that all conclude the same thing: project completion rates are abysmal! So, the question we pose to you is this. What if we could demonstrate a different method that would push your project successful completion rate from where your rate is now to over 90 to 95 percent? Would you be interested in hearing about it?

Ninety percent of the project managers around the world are using a project management method referred to as the Critical Path Method and have been doing so for many years. If you ask a typical project manager about what factors delayed a completed project, most will tell you that something they didn't expect or even had no control over cropped up in some of the tasks and delayed them. In other words, uncertainty or the Murphy bug bit them! Every project from virtually every environment has uncertainty associated with it, and how this uncertainty is dealt with determines the ultimate success or failure of the project. So, for a project to be successful, doesn't it make sense that there must be a way to protect it from uncertainty? Let's take a look at how traditional project management (CPM) protects a project from uncertainty.

CPM uses a "fudge factor" to protect projects from inevitable uncertainty. That is, when developing the project plan, durations for each individual task are estimated by the resources responsible for executing them, and then a safety factor is added to each of the tasks by the resource responsible for completing them. For example, suppose the realistic estimate of time for an individual task is one week. Is one week what the resource actually tells his or her project manager? Typically, the resource will add a safety factor of their own, to guard against "things" that might happen to cause a delay in the completion of the task. So, it's not unusual for the original one week to be quoted as two weeks. Resources react this way because they know from experience that as soon as they give the project manager an estimate, it automatically becomes a commitment!

A typical project manager will add up all of the individual, inflated time estimates and then add his or her own safety factor. Why is this done? Project managers know that at some point in the project Murphy will strike, and some of the tasks will be delayed, so they too add a safety factor to protect the project from being late. Keep in mind that every resource inflates every task, so it's not uncommon for the estimated duration to be more than 50 percent longer than it takes to actually complete the task. So with all of this built-in safety, the project should be completed on time, right? So it seems, but the statistics on project completions don't bear this out. We'll explain why later.

In traditional project management, how do you track progress against the completion date? The typical method involves calculating the percentage of individual tasks completed and then comparing that percentage against the due date. Sounds reasonable, but is this the right way to track progress? The problem with using percentage of tasks completed is that not all tasks have the same duration. That is, comparing a task that has an estimate of one day to a task that should take one week is not a valid comparison. Compounding this problem is the mistaken belief that the best way to ensure that a project will finish on time is to try to make every individual task finish on time. This too sounds reasonable, but later on we'll show you why this isn't so.

If individual project tasks have so much extra time imbedded in them, then why are so many projects coming in late? We think that this is partially explained by two common human behaviors. Resources know that they have built "safety" into their tasks, so they often delay work on the task until much later than they had originally planned. Think back to your high school days. When you were given a due date for a paper for say, next Thursday, when did you actually start working on it? Wednesday? Eli Goldratt coined the term the Student Syndrome to explain one of the reasons why the apparent built-in safety gets wasted. When the task start is delayed and Murphy strikes, the task will be late.

Another human behavior that lengthens projects is referred to as Parkinson's Law. Resources intuitively know that if they finish a task in less time than they estimated, the next time they have the same or a similar task to complete, they will be expected to finish it early also. So, to protect against this, even though the task is finished early, the resource doesn't notify the project manager that it is finished until the original due date. After all, we're talking about credibility here, so to protect their credibility, early finishes aren't reported. Parkinson's Law states that work expands to fill the available time, so resources use all of the available time. The key effect on projects of these two behaviors is that delays are passed on, but early finishes aren't. So, is it any wonder that projects are late?

While these two behaviors negatively impact project schedules, there are other reasons why projects are late. Many organizations today have multiple projects going on at the same time, and it's not unusual for projects to share resources. In fact, many project managers tend to "fight over" shared resources, because they believe their project is the one that has the highest priority. Another significant problem is that in many project-based companies, leadership initiates projects without considering the capacity of the organization as a whole to complete the work. Leadership often mistakenly assumes that the sooner a project is initiated, the sooner it will be completed. As a result, perhaps the most devastating problem of all associated with project completion occurs, which is multi-tasking! But wait a minute, I thought we'd all been taught for years that multi-tasking is a good thing?

Multi-tasking happens when resources are forced to work on multiple project activities at the same time. Like we've always said, humans aren't very good at rubbing their tummy and patting their head at the same time. Many people believe, especially in leadership positions, that multi-tasking is a good thing because it increases efficiency, since everyone is "busy" all of the time. If you've ever to read *The Goal* by Eli Goldratt (if you haven't, you should), you might remember how focusing on local activities actually damaged the overall system performance. You may also recall how Goldratt used his robot example, whereby running the robots continuously did improve efficiency, but at the expense of creating excessive amounts of work-in-process (WIP) and finished goods inventory.

The negative impact of multi-tasking in a project management environment is much, much worse. Let's look at an example. Per Figure 7.1, suppose you have three projects that you are assigned to, and in each project you have estimated that you have two weeks (ten days) of work on each project for the tasks assigned to you. Assuming Murphy didn't strike, if you started and finished Project 1 without stopping or working on any other project, it would be done in ten days. Ten days because that's what you told everyone it would take (Parkinson's Law). Likewise, for Projects 2 and 3, assuming no other interruptions, each would take ten days to complete for a total time to complete the three projects of 30 days. But having laid it out

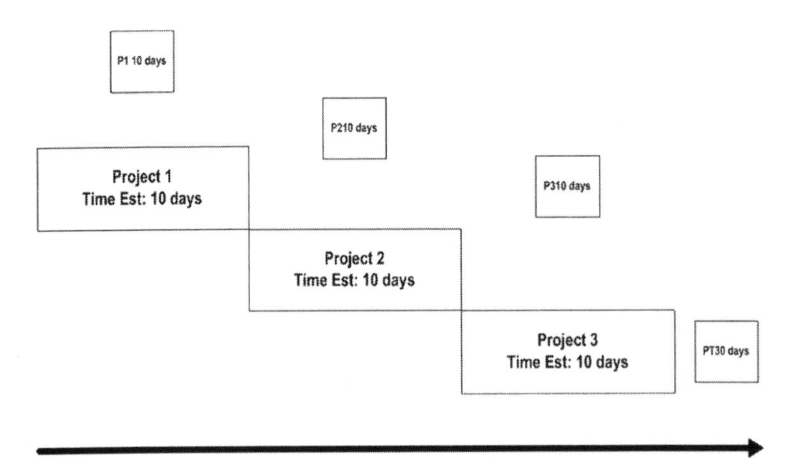

FIGURE 7.1
Example of Three Projects Selected.

like this, if all three projects were scheduled to start on the same day, then Project 1 would be on time at ten days, Project 2 would be done in 20 days (i.e. ten days late) and Project 3 would be done in 30 days (i.e. 20 days late).

But can you reasonably expect to work on only one project at a time? Because there are probably three different project managers (i.e. one for each individual project), each one is most likely telling you (or maybe even screaming at you) that they need you to show progress on their project (as you will see later on, CPM reports project progress typically by measuring percentage of tasks complete). And because you want to satisfy all three managers, you decide to split your time between the three projects (i.e. you will multi-task). So as is demonstrated in the Figure 7.2, you start with Project 1 and work on it for three days. On the fourth working day, you stop working on Project 1 and begin Project 2 and work on only it for three days. On the seventh working day, you stop working on Project 2 and begin working on Project 3. On the tenth day, you stop working on Project 3 and begin working on Project 1 again. You repeat this sequence until all three projects are completed as in Figure 7.2.

By multi-tasking, look what's happened to the time to complete each individual project. Without multi-tasking, each individual project took only ten days to complete and 30 days to complete all three. *Without* multi-tasking, Project 1 was completed in ten days, Project 2 in 20 days, and

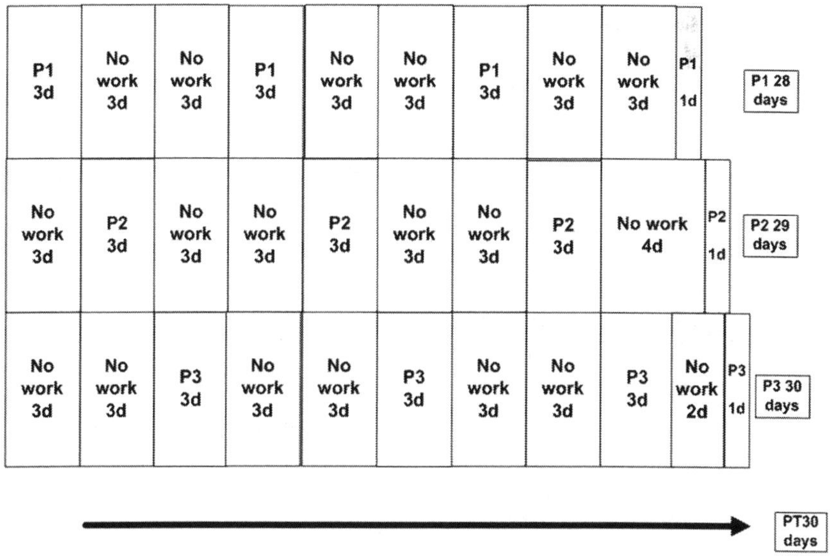

P1 3d	No work 3d	No work 3d	P1 3d	No work 3d	No work 3d	P1 3d	No work 3d	No work 3d	P1 1d	P1 28 days	
No work 3d	P2 3d	No work 3d	No work 3d	P2 3d	No work 3d	No work 3d	P2 3d	No work 4d	P2 1d	P2 29 days	
No work 3d	No work 3d	P3 3d	No work 3d	No work 3d	P3 3d	No work 3d	No work 3d	P3 3d	No work 2d	P3 1d	P3 30 days

PT30 days

FIGURE 7.2
Completed Projects.

Project 3 in 30 days. *With* multi-tasking, Project 1 took 28 days, Project 2 took 29 days and Project 3 took 30 days. Both methods completed all three projects in 30 days, but which set of results do you think your leadership would prefer? Also, keep in mind that when you exercise multi-tasking, there is also time required to get reacquainted with each project, so the multi-tasking times will actually be considerably longer. In fact, some studies have shown that tasks often take two to three times their estimated duration when multi-tasking occurs.

So, let's summarize what we've learned before we move on. We've learned that task time estimates for tasks are artificially lengthened as a protective measure against Murphy and all of the negative baggage he brings to the party. We've learned that even though this safety is built in, it is wasted because of the Student Syndrome and Parkinson's Law. With the Student Syndrome, we put off work (i.e. procrastinate) until the last minute, while Parkinson's Law says that if we're given ten days to complete a task, that's how long it will take, even if it is completed earlier. And finally, we've learned how devastating multi-tasking can be to the completion rate of projects. So, what can we do about these three behavioral problems? What if we could significantly reduce these imbedded safety buffers and still provide the protection that we need?

A DIFFERENT METHOD

As we've seen, in CPM task durations are inflated to protect against Murphy's untimely attacks. What if we could significantly reduce these imbedded safety buffers and still provide the protection that we need? In our earlier example, suppose we were able to reduce the estimated duration by 50 percent and still protect against Murphy. In other words, if we could complete the tasks in five days instead of ten days, wouldn't this be a quantum leap in project completion time reduction?

Figure 7.3 is a graphical depiction of the reduction in durations of each project. We have just reduced the time to complete these three projects from 30 to 15 days, but can we do this and safely guard against the uncertainty introduced by Murphy? The answer is, yes we can! But before we explain how we can do this, we need to re-introduce some of you to something called the Theory of Constraints.

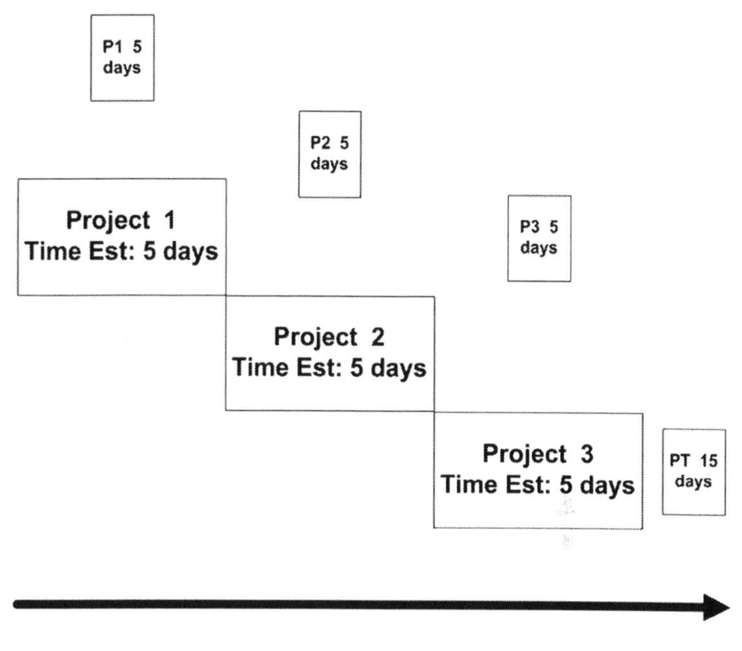

FIGURE 7.3
Graphical Depiction of the Reduction in Durations of Each Project.

TOC came on the scene in the mid-1980s through its developer, Eli Goldratt. Goldratt taught the world that every organization has at least one (and usually only one) constraint that prevents an organization from coming closer to its goal. And for most companies, the goal is to make money now and in the future. In fact, Goldratt analogized this concept to the strength of a chain being dictated by its weakest link.

In a manufacturing environment, TOC presumes that only one workstation, the one with the least capacity, dictates the throughput of the process, and to operate all stations at maximum capacity will only serve to create excess inventory in the process. This excess inventory increases the lead time and wastes resources. The best way to understand TOC is to envision a simple, repeatable process as in Figure 7.4.

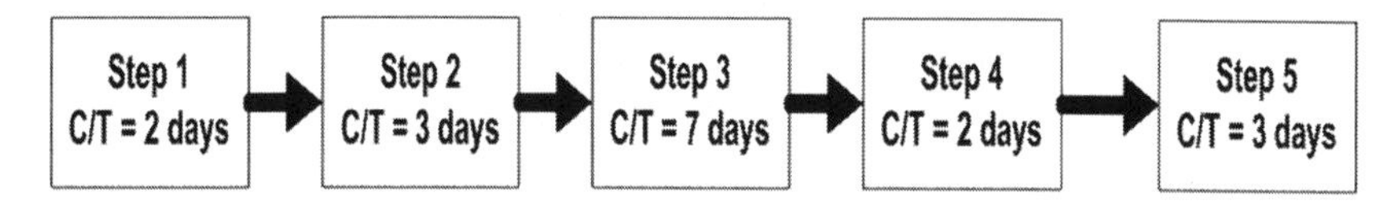

FIGURE 7.4
Example of a 5-Step Process.

In Figure 7.4, Step 3 at seven days is the slowest resource since it requires seven days to complete it, whereas all of the others require much less than seven days. Therefore, the maximum throughput of the process in Figure 7.4 is one unit of product (or service) every seven days. TOC identifies Step 3 as the constraint and tells us that if we want to improve throughput, then we must focus our improvement efforts on this step.

The question you might have in your mind right now is, why bring up the Theory of Constraints when we're talking about project management? The answer is actually quite simple. TOC recognizes the existence of a constraint, and this recognition in a project management environment is absolutely critical to shortening the time required to complete projects.

TOC teaches us that the first order of business is to define your Goal. In a single project, the goal is identified as making the promised project due date, while in a multiple project environment, the goal is to maximize the throughput of projects. So how do Goldratt's 5 Focusing Steps apply to project management?

Goldratt's 5 Focusing Steps:

1. Identify the system constraint.
2. Decide how to exploit the constraint.
3. Subordinate everything else to the constraint.
4. If necessary, elevate the constraint.
5. Return to Step 1, but don't let inertia create a new constraint.

Good question, and in my next section, I will demonstrate how each of the five steps applies to a project management environment, as well as introducing TOC's version of project management, CCPM.

CRITICAL CHAIN PROJECT MANAGEMENT

Earlier we demonstrated how, by simply eliminating multi-tasking, significant gains can be made in project completion rates, but we still have to address the impact of the Student Syndrome and Parkinson's Law. We know that both of these behaviors work to lengthen the time required to complete projects. Remember how excess safety is imbedded into traditional project management plans? Resources estimate task times and add in their own protection against disruptions caused primarily by

Murphy. Knowing that this safety exists, resources then delay starting work on their tasks until the due date is close. Even if the resources don't delay the task starts and finish early, these early finishes are not reported and passed on. So how does CCPM deal with these two behaviors?

While CPM relies on individual task durations, as well as scheduled start and completion dates, CCPM does not. The focus is no longer on finishing individual tasks on time, but rather is on starting and completing these tasks as soon as possible. So how does this work? Like CPM, CCPM still gathers estimates on individual tasks and identifies its own version of the Critical Path. Unlike CPM, CCPM considers competing resources (i.e. the same resource has to work on different tasks) and makes them a part of the critical path. Let's look at an example of how CPM and CCPM identifies the critical path.

CPM defines the critical path as the longest path of dependent tasks within a project. That is, tasks are dependent when the completion of one task isn't possible until completion of a preceding task. The critical path is important, because any delay on the critical path will delay the project correspondingly. Figure 7.5 is an example of a series of tasks that must be completed in a project, with the critical path highlighted in grey. Traditional project management determines the critical path by looking at the task dependencies within the project. Task A2 can only be initiated after A1 is completed. Task B3 can only be performed after completion of B2, and C2 only after C1. Task D1 can only be performed after completion of A2, B3 and C2. Using CPM, the critical path would have been identified as C1-C2-D1, and the project completion estimate would have been 29 days (i.e. 8d + 12d + 9d).

In addition to task dependencies, there are also resource dependencies that CPM fails to recognize. What if, in our example, tasks A2 and B3 are performed by the same resource? Is the critical path different? In Figure 7.6, we see the new critical path that includes a provision for resource dependencies, and as you can see the new critical path is 5d + 10d + 10d + 9d, or 34 days. So, the minimum time to complete this project is now 34 days.

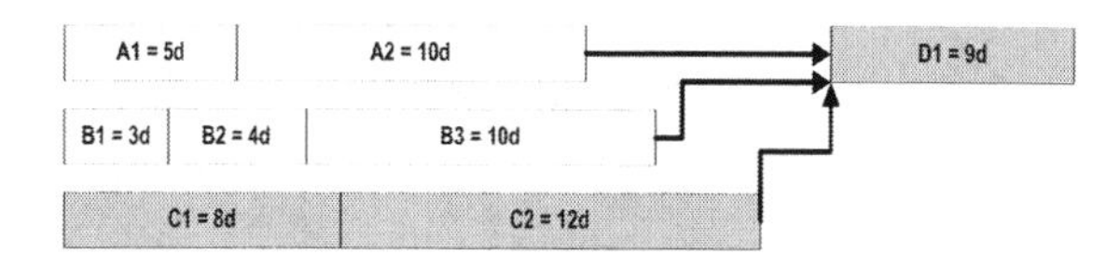

FIGURE 7.5
Example of a Series of Tasks Which Must Be Completed in a Project.

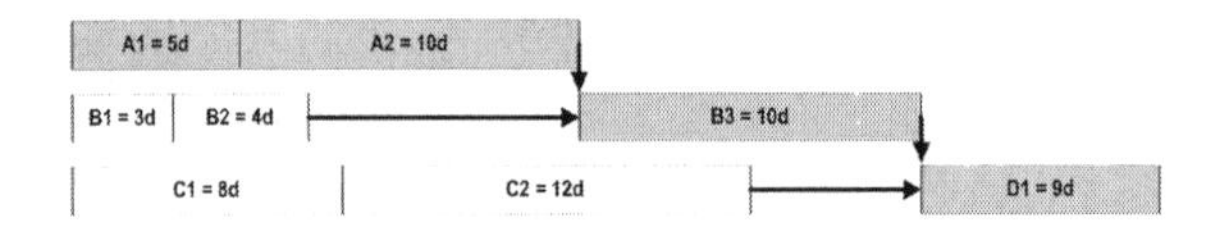

FIGURE 7.6
The New Critical Path That Includes a Provision for Resource Dependencies.

In our opinion, the failure to consider resource dependencies is one of the key reasons why project completion rates are so terrible. The simple implication of incorrectly identifying the critical path, which we will now refer to as the *Critical Chain*, is that the project team will never be able to complete their project on time without heroic efforts, adding additional resources, overtime or a combination of all three. The practical implication of incorrectly identifying the real Critical Chain is that the focus will be on the wrong tasks. Is this any different than focusing on non-constraints in our earlier discussion on TOC?

We said earlier that safety is imbedded within each task, as a way to guard against the uncertainties of Murphy. Critical Chain takes a completely different approach, by assuming that Murphy's uncertainty will happen in every project. Unlike CPM, CCPM removes these safety buffers within each task and pools them at the end of the project plan. In doing so, this protects the only date that really matters—the project completion date. There are many references that explain the details of how CCPM does this, but here's a simple example to explain it. Basically, we have removed all of the protection from individual task estimates, which we estimate to be 50 percent of the original estimate. Figure 7.7 demonstrates the removal of this safety. So now, the length of the Critical Chain is no longer 34 days, but rather 17 days. But instead of just eliminating the safety buffer, we want to place it where it will do the most good, at the end of the project to protect the project due date. This isn't exactly how this works, but for presentation purposes to demonstrate the theory behind CCPM, it will suffice.

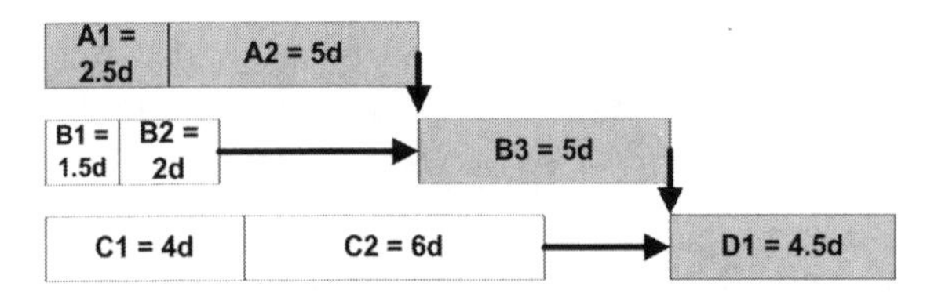

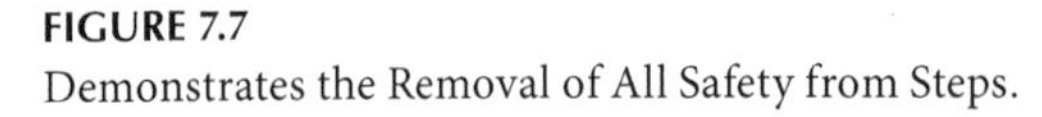
FIGURE 7.7
Demonstrates the Removal of All Safety from Steps.

Figure 7.8 is this same process, but this time the safeties that we removed are added to the end of the project to act as a cushion against Murphy's inevitable attack. So, the question now becomes, how do we utilize this buffer, and how does it improve the on-time completion of the project?

Suppose task A2 takes seven days, instead of the five days that are in the plan? In a traditional project management environment, this would be cause for panic. In a CCPM environment, we simply consume (borrow) two days from the project buffer, and we're still on schedule. Suppose now, for task B3, we only take three days instead of the planned five days. We simply add the gain of two days, back into the project buffer. In traditional CPM, delays accumulate while any gains are lost, and this is a significant difference! The project buffer protects us from delays. For non-Critical Chain tasks, or subordinate chains such as C1-C2 from our example, we also can add feeding buffers to assure that they are completed prior to negatively impacting/delaying the Critical Chain.

One of the key differences between CPM and CCPM is what happens at the task level. In traditional project management, each task has a scheduled start and completion date. CCPM eliminates the times and dates from the schedule and instead focuses on passing on tasks as soon as they are completed, much like a runner in a relay race passing the baton. This function serves to eliminate the negative effects of both the Student Syndrome and Parkinson's Law from the equation and permits on-time and early finishes for projects. For this to work effectively, there must be a way to alert the next resource to get ready in time.

Earlier, we explained that in traditional project management, we track the progress of the project by calculating the percentage of individual tasks completed and then comparing that percentage against the due date. The problem with this method is that it is nearly impossible to know exactly how much time is remaining to complete the project. Using this method to track progress, often you'll see 90 percent of a project completed, only to see the remaining 10 percent take just as long. In fact, looking at the number or percentage of tasks completed instead of how much of the critical path

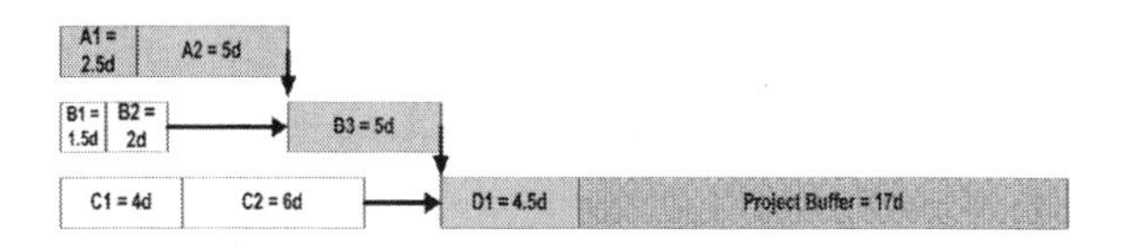

FIGURE 7.8
The Safeties That Were Removed Are Added to the End of the Project in Figure 7.7.

has been completed only serves to give a false sense of conformance to the schedule.

CCPM measures the progress of a project much differently and in so doing allows the project to make valuable use of early finishes. Critical Chain uses something called a *Fever Chart,* which is simply a run chart of percentage of the Critical Chain complete versus percentage of Project Buffer consumed. Figure 7.9 is an example of such a chart. In this chart, we see that approximately 55 percent of the Critical Chain has been completed, while only 40 percent of the project buffer has been consumed. At this rate, we can conclude that this project is slightly ahead of schedule.

The green, yellow and red areas of the fever chart are visual indicators of how the project is progressing. If the data point falls within the green area of the chart, the project is progressing well and may even finish early. If the data point falls into the yellow zone there is cause for concern, and plans should be developed to take action if necessary. Vertical rises indicate that buffer is being consumed at too fast a rate relative to how the project is progressing. If a data point falls into the red zone, then the plan we developed must now be executed. But even if the entire amount of project buffer is consumed at the completion of the project, the project is still on time and not late.

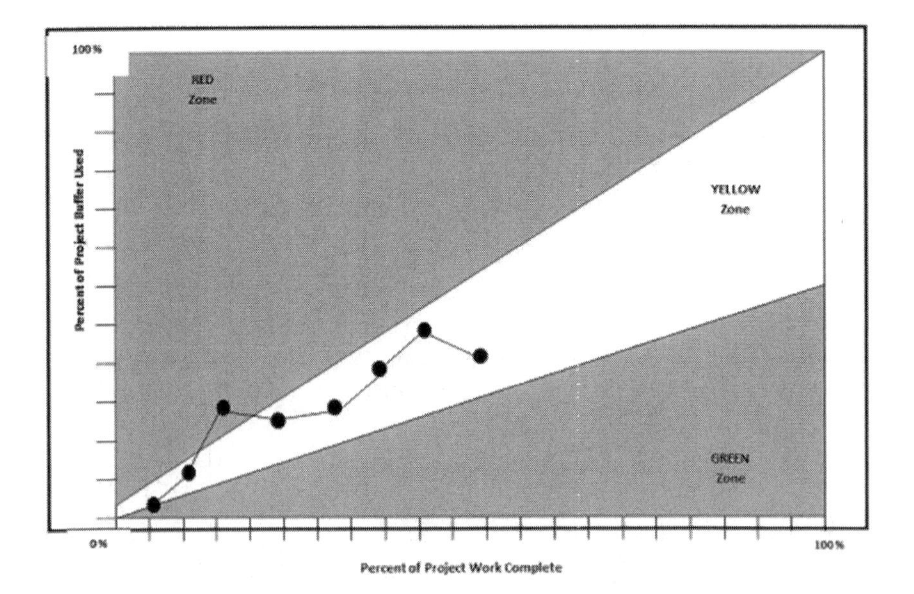

FIGURE 7.9
Example of a Fever Chart.

In addition to using the fever chart, we also recommend calculating a project index by dividing the percent of Critical Chain completed into the percent of the project buffer consumed. As long as this ratio is 1.0 or less, then the project will come in on-time or early. In our example, this ratio would be 40 percent divided by 55 percent, or 0.727. This ratio says that this project is progressing nicely with no concern for the project being late.

With most CCPM software, we can also see a portfolio view of the fever chart that tells us the real-time status of all projects in the system. Figure 7.10 is an example of this view, and one can see at a glance that four of the projects (Projects 1, 4, 5 and 6) need immediate attention (they are in the red zone), two projects (Projects 3 and 8) need a plan developed to reduce the rate of buffer consumption (yellow zone) and two projects (Projects 2 and 7) are progressing nicely (in the green zone). Having this view enables the project manager to see at a glance where to focus his or her efforts. It is important to understand that just because a project enters the red zone does not mean that the project will automatically be late. It only means that if prompt action isn't taken to reduce the buffer consumption rate, the project could be late.

The net effect of CCPM will always result in a significant decrease in cycle time, with a corresponding increase in the throughput rate of completed projects using the same level of resources. In fact, it is not unusual for

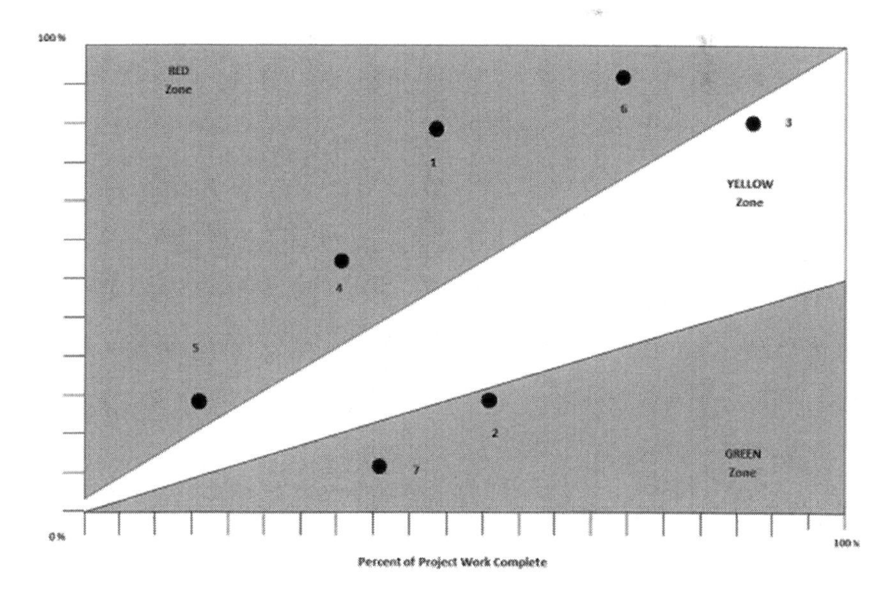

FIGURE 7.10
Example of Real-Time Status of All Projects in the System.

project cycle times to be reduced by as much as 40 percent to 50 percent! These cycle time reductions and throughput increases translate directly into improved on-time delivery of projects as well as significant revenue increases.

The key to success using CCPM revolves around utilization of the true subject matter experts (SMEs). That is, by developing a *core team* comprised of 70 percent to 80 percent employees actually executing projects (i.e. SMEs) and permitting them to develop the ultimate design solution, the resulting implementation will be believed, owned and used by the people performing the work. This ownership translates directly into making sure the solution will work. Without this level of involvement and approval authority to develop the applied action plans, CCPM will simply not be as successful as it could or should be.

Another key to successful project management application is a series of regular meetings intended to escalate and resolve any problems that surface during the project execution. These include daily "walk-arounds" by the project manager with the SMEs to determine the progress of the project so that problems can be surfaced and escalated if the project manager cannot resolve them him or herself. In addition, I also recommend that each week or two there must be what is referred to as an Operations Review in which each individual project is reviewed for progress by the leadership team. Again, if problems and issues need to be escalated to keep the project on schedule, leadership must play this vital role.

Clearly, Critical Chain Project Management has demonstrated its superiority over the predominant Critical Path Method in a variety of industry settings, with reported cycle time improvements in the neighborhood of 40 percent to 50 percent. And with project completion rates above 90 percent, it is no wonder why organizations that rely on project completions as their revenue source are flocking to CCPM. Organizations like the Department of Defense (DoD), where rapid maintenance of aircraft and other military vehicles is paramount to success, and software development companies have had incredible success using CCPM versus CPM. So, if your organization is a project-based one with multiple projects, CCPM will take you to a new level of success.

Earlier, we told you we would tie Goldratt's 5 Focusing Steps into Critical Chain Project Management and provide a summary of what we've written on this subject. Let us now refresh your memory on Goldratt's focusing steps, while simultaneously tying them into Critical Chain Project Management:

1. *Identify the system's constraints*: For a single-project environment, this simply means that identifying the *Critical Chain* or the longest chain or longest path of dependent tasks within a project determines the actual duration of the project. The Critical Chain is therefore the constraint. In a multi-project scenario, there is a drum resource that limits the number of projects that an organization can manage and deliver. This resource, more than any other, controls the flow of projects and is considered the constraint.
2. *Decide how to exploit the constraint*: For a single-project scenario, this simply means focusing on the Critical Chain tasks, to make sure that the required work is done without unnecessary delays. In a multi-project situation, this means that projects should be prioritized and then staggered according to the drum resource's capacity, making sure it is not overloaded.
3. *Subordinate everything else to the previous decision*: As you might have concluded, this simply means that non-Critical Chain tasks cannot, and must not, interfere with or delay work on the Critical Chain. To avoid this scenario, we have strategically placed feeding buffers to prevent delays on the Critical Chain. In multi-project situations, non-critical resources may have to wait in favor of the Critical Chain resources.
4. *Elevate the system's constraint*: For single- and multi-project environments, this typically means investing in additional resources or even increasing the capacity of resources that impact both the Critical Chain or project throughput. Often, this might mean spending money or using non-critical resources to support Critical Chain tasks.
5. *Return to Step 1*: When one project is completed, identify/insert the next one and proceed to Step 2.

SUMMARY OF KEY POINTS

- In a fairly recent survey (The Chaos Report), by the Standish Group, studying nearly 10,000 IT projects across America, it was reported that 52 percent of projects ended up costing greater than 189 percent of the original budget, 31 percent were cancelled and only 16 percent of the projects were completed on time and on budget. The fact is, there are many other reports from numerous industry types, from all over the world, that all conclude the same thing—project completion rates are abysmal!

- Ninety percent of the project managers around the world are using a project management method called the *Critical Path Method* and have been doing so for many years. CPM uses a "fudge factor" to protect projects from inevitable uncertainty. That is, when developing the project plan, durations for each individual task are estimated by the resources responsible for executing them, and then a safety factor is added to each of the tasks by the resource responsible for completing them. In *Critical Chain Project Management*, individual tasks durations are removed and replaced with a project buffer.
- In traditional project management (CPM), tracking is done by calculating the percentage of individual tasks completed, and then comparing that percentage against the due date. CCPM tracks progress on the Critical Chain against buffer consumption.
- There are behavioral issues associated with traditional project management (CPM). These issues are the *Student Syndrome* (or procrastinating start of the project because of the built-in safety buffers); *Parkinson's Law* (work expands to fill the available time); and *Multi-tasking* (moving back and forth between multiple projects thus extending the duration of all of the projects). CCPM eliminates these behavioral issues by eliminating individual task durations using the relay runner scenario (i.e. passing on a task as soon as it is completed), and staggering or pipelining the projects (i.e. delaying project starts).

Whereas CPM completion rates are clearly abysmal, completion rates using CCPM are excellent (i.e. typically >90 percent to 95 percent), and the completion times are usually 40 percent to 50 percent faster. In addition, when comparing scope and cost, surveys of companies using CCPM show that CCPM is a far superior project management method.

8

Theory of Constraints Replenishment Solution

INTRODUCTION

In many manufacturing facilities and even Maintenance, Repair and Overhaul (MRO) facilities, the system typically in use to obtain parts and supplies is referred to as a Min/Max replenishment system. If you're in the defense industry (i.e. Department of Defense), then we're sure this is the system being used. One thing that is certain about these type systems is that they will eventually stop working! And often, the Purchasing Manager is actually measured on the basis of how much money can be saved. That is, the more money saved, the bigger his or her incentive bonus will be. Consequently, when you combine both of these factors, it's typically a recipe for disaster in terms of parts availability. Think about it: on the one hand you need the parts to produce whatever it is you produce, while on the other hand you have the group responsible for keeping an adequate parts supply working to minimize the parts inventory. These two factors seem to be in conflict with each other, don't they? So how do we resolve this conflict? Let's take a look.

The Min/Max supply system has three basic rules that must be followed:

- Rule 1: Determine the minimum and maximum stock levels for each part.
- Rule 2: Don't exceed the maximum stock level for each part.
- Rule 3: Don't re-order until you reach or go below the minimum stock level for each part.

The driving force behind these Min/Max rules is deeply imbedded in the cost world belief that in order to save money, you must reduce the amount of money being spent on parts. To do this, you must never buy more than the maximum level and never order until you reach or go below the minimum level. It's the age-old conflict of *saving money* versus *making money*.

The theory behind this Min/Max concept assumes that parts are stored at the lowest possible level of the supply chain, usually at the point of use and usually in a parts bin. The parts are then used until the calculated minimum quantity is either met or exceeded. When the minimum quantity is met or exceeded, an order is placed to replenish the parts back to the maximum level. The parts order proceeds up the chain from the bin they're kept in to the production supply room and then on to the central warehouse where they are ordered.

Figure 8.1 is a visual flow of what I just described, and as you can see, the distribution of parts is from the top-down and the re-order is from the bottom-up. The parts come into the central warehouse from the suppliers, and from there they are distributed to your plant stock room. The parts are distributed to the appropriate line stock bins until they are needed in your operations. Typically, once a week the bins are checked to determine the inventory level in each of the bin boxes. If the bins are at or below the minimum defined level, then an order is placed for that part number. Even though this type of system "appears" to control the supply needs of your plant, in reality there are negative effects that we see and feel with the Min/Max system.

The real danger in using the min/max system is pointed out in Figure 8.1. Because parts are not re-ordered until the re-order point is reached or surpassed, there is a high likelihood that stock-outs will occur for potentially extended periods of time (Figure 8.2).

This problem of stock-outs seems to occur over and over again, as depicted in Figure 8.3. So, what's the solution to this stock-out dilemma?

As stated earlier, even though the Min/Max type of system "appears" to control the supply needs of your plant, in reality there are some very negative effects that we see and feel with it. The first problem we experience with this Min/Max system is that you are continuously in a reactive and knee-jerk state, rather than a proactive and practical mode. This is simply because the Min/Max system is almost always assured to have "stock-out" conditions regularly and repeatedly. So, the question we must answer is why do these stock-outs occur, and what can we do to prevent them?

Very simply put, stock-outs occur principally because it's not unusual for the lead time to replenish the minimum amount of parts left in the bin,

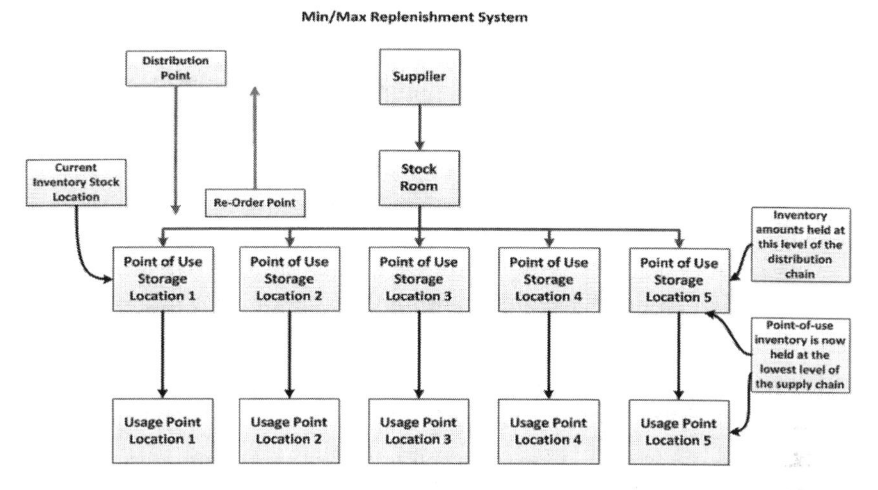

FIGURE 8.1
Visual Flow of Parts in a Process.

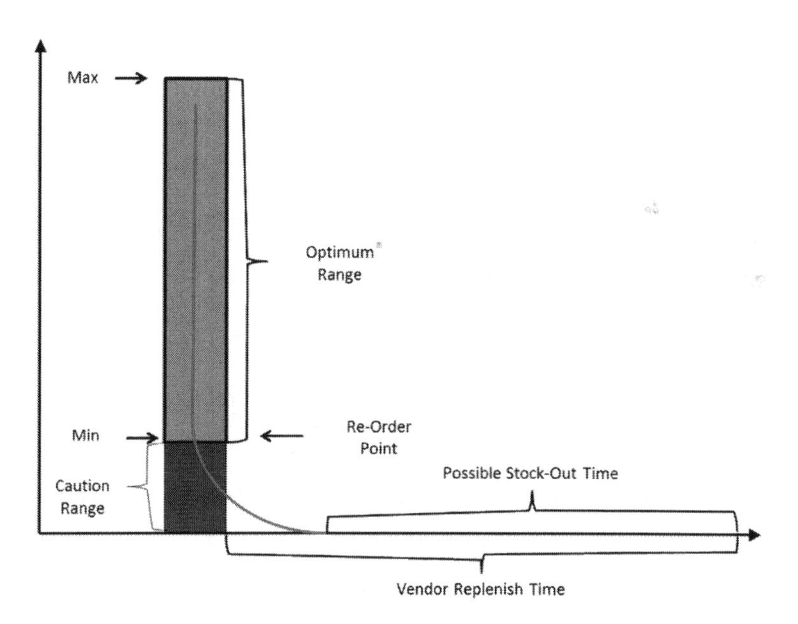

FIGURE 8.2
The Real Danger in Using the Min/Max System.

quite regularly exceeds the time remaining to build products with what's left in the part bin. And because of the inherent variation in demand, stock-outs can occur in both shorter or longer times than the Min/Max model might propose. The problem with this is that when you do have

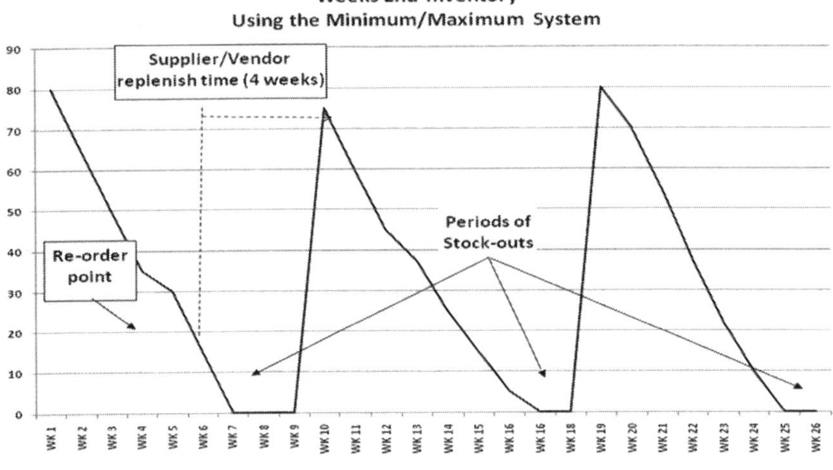

FIGURE 8.3
Stock-Outs Occurring Over and Over Again.

a stock-out, your production stops until new parts arrive, and it usually happens often.

As mentioned earlier, the flow in the Min/Max Parts Supply System, where the parts are distributed to the lowest level of the distribution (i.e. to parts bins) and are also re-ordered from this same low point in the system. So, what must be done differently to avoid these stock-out periods? Wouldn't it be great if we had a system that would operate with much less on-hand inventory without stock-outs?

THEORY OF CONSTRAINTS REPLENISHMENT MODEL

There is a system, and it comes to us from the Theory of Constraints (TOC). The TOC Distribution and Replenishment Model states that, unlike the Min/Max system, most of the inventory should be held at the highest level in the distribution chain and not at the lowest level (i.e. the bins). Of course, you must hold some inventory at the point of use (POU) for your assembly work, but this model tells us that the majority of it should be held at the warehouse, from where it's ordered and received from the supplier. The bottom line is this: instead of using some minimum quantity to trigger the re-order of parts, the re-order process should be triggered by daily usage and the time required for the vendor to replenish the parts.

That is, it tells us to simply replace what we've used on a very frequent basis, rather than waiting for some minimum quantity to be reached. When this system is used, there will always be enough parts on hand to produce your products, and no stock-outs will occur! Figure 8.4 is a graphic image of what the TOC Replenishment System might look like.

Buffers are placed at better leverage points in the supply chain (i.e. most inventory is held in the factory warehouse). Likewise, each regional warehouse and retail location has buffers for each product. These buffers (physical products) are divided into green, yellow and red zones and are located at strategic locations to avoid stock-outs. This replenishment system relies on *aggregation* to smooth demand, and demand at regional warehouses is smoother than demand at retail locations, simply because higher-than-normal demand at some retail locations is off-set by lower demand at other ones.

Demand at the factory warehouse is even smoother than demand at the regional warehouses. Goods produced by the factory are stored in a nearby warehouse until they are needed to replenish goods consumed by sales.

Because sales occur daily, shipments occur daily, and the quantities shipped are just sufficient to replace goods sold. This might seem to increase shipping costs over what could be achieved by shipping large batches less frequently, but the truth is, the net effect on total shipping costs is that they actually decrease. Stopping the shipment of obsolete

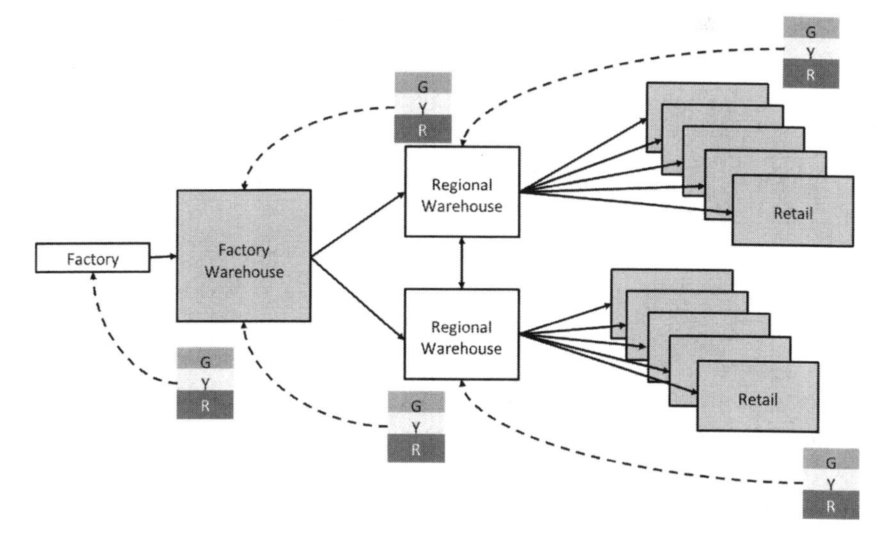

FIGURE 8.4
Graphic Image of What the TOC Replenishment System Might Look Like.

goods and reshipment of misallocated goods more than compensates for increased costs created by smaller shipments of saleable goods.

The ability to capture sales that would otherwise be lost due to insufficient inventory makes the TOC solution a better alternative. In this system, replenishment is driven by actual consumption and not a sales forecast. As sales are made, the buffer levels at retail locations drop, eventually triggering replenishment from the factory warehouse, which triggers a manufacturing order to resupply the appropriate buffer before it runs out.

Buffer sizing is based on both variability and the time it takes to re-supply. So, the more variable the consumption is, the larger the buffer must be to cover the variability. In addition, the longer it takes to re-supply, the bigger the buffer needs to be, to be able to cover the demand during the re-supply waiting times.

The benefits of TOC's replenishment solution can be very striking. For example, a traditional distributor that is 85 percent reliable can reasonably expect to increase its reliability to 99 percent while reducing its inventory by two-thirds. In addition, the average time to resupply retail locations typically drops from weeks or months to about one day. A central benefit of TOC's replenishment solution is to change the distribution from push to pull. That is, nothing gets distributed unless there is a market for it. Market pull, the external constraint, then optimizes distribution while minimizing inventory.

As an added bonus for using this system, the average overall inventory will be significantly lower. This happens because when an order is placed under the Min/Max system, the system automatically re-orders, but does so to the maximum quantity. Since the TOC Replenishment System simply re-orders what's been used, the amount of inventory required to be on hand drops significantly. In fact, you'll see a drop in inventory levels in the neighborhood of 4 percent to 60 percent! Imagine what that means to your cash flow. Figure 8.5 is a graphical summary of what you can expect.

A simple way to present the TOC Replenishment System is by considering a soda vending machine. When the supplier (the soda vendor) opens the door on a vending machine, it is very easy to see what products have been sold. The soda person knows immediately which inventory has to be replaced and to what level to replace it. The soda person is holding the inventory at the next highest level, which is on the soda truck, so it's easy to make the required distribution when needed. The soda person doesn't leave six cases of soda when only 20 cans are needed. If he were to do that,

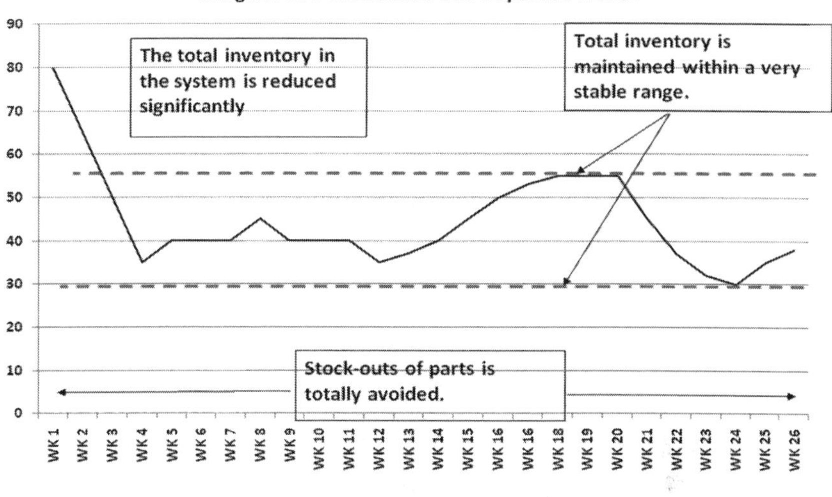

FIGURE 8.5
Graphical Summary of What You Can Expect with TOC Replenishment System.

when he got to the next vending machine he might have run out of the necessary soda, because he made distribution too early at the last stop.

After completing the daily refill of the vending machines, the soda person then returns to the warehouse or distribution point to replenish the supply on the soda truck and get ready for the next day's distribution. When the warehouse makes distribution to the soda truck, they move up one level in the chain and replenish what's been used from their supplier. Replenishing in this way significantly reduces the on-hand inventory, while significantly reducing stock-outs. If a type of soda always runs out, then more should be added to the vending machine (i.e. another row of soda).

So, to summarize the differences:

- The TOC Replenishment Model holds a small amount of inventory at the (POU), while holding the majority of the inventory at the highest level of the organization, typically in a central warehouse. The Min/Max system holds all of the inventory at the POU.
- The TOC Replenishment Model re-orders parts based upon real usage on a frequent basis (i.e. typically weekly) and orders parts from the central warehouse. The Min/Max system re-orders to the maximum level, when the number of parts remaining in the parts bin meets or goes below the calculated minimum quantity, and then

orders directly from the POU. Often, the time required to replenish the part exceeds what's left in the parts bin, and a stock-out occurs.
- Because the TOC Replenishment Model re-orders what's been used on a frequent basis, no stock-outs occur, and typically the level of inventory is reduced by approximately 50 percent.

SUMMARY OF BENEFITS FROM TOC'S REPLENISHMENT SYSTEM

1. The reduction of total inventory required to manage and maintain the total supply-chain system is typically in the order of 40 percent to 60 percent, while experiencing the virtual elimination of stock keeping units (SKU) stock-out situations.
2. Distribution of SKU is made at the right time, to the right location.
3. The frustration caused by stock-out situations virtually disappears—not only in being able to complete the work, but also the elimination in the time spent looking for and waiting for SKUs to become available.

Because waiting due to stock-outs virtually disappears, parts flow and synchronization improves dramatically, which improves the throughput of parts through the entire supply chain. And because throughput improves, profitability increases proportionally to the level of sales.

9

Understanding Variation

INTRODUCTION

One of the keys to optimizing the performance of your processes is understanding the nature and sources of variability. Like anything else, if you don't understand it, you simply won't be able to improve it. In this chapter, we are going to discuss a series of key points related to process and system variability.

Hoop and Spearman [1], in their breakthrough book, *Factory Physics*, provide valuable insights into variability by providing us with seven fundamental points to remember. These seven points are as follows:

1. *Variability is a fact of life.* From a management perspective, it is clear that the ability to recognize and deal effectively with variability is perhaps the most critical skill to develop for all Managers, Engineers and Black Belts. Without this skill, your decisions will be full of uncertainty and might even be wrong many times.
2. *There are many sources of variability in manufacturing systems.* Process variability comes at us in many different forms. It can be as simple as work method variations, or as complex as machine setups and changeovers, planned and unplanned downtime or scrap and rework. Flow variability is created by the way we release work into the system or how it moves between stations. The result of variability present in a system can be catastrophic if its underlying causes aren't identified and controlled.
3. *The coefficient of variation is a key measure of item variability.* The coefficient of variation, given by the formula $CV = \sigma/\mu$, is a reasonable way to compare the variability of different elements of a production or flow system. Because it is a unitless ratio, we can make rational comparisons of the level of variability in both process times and

flows. In workstations, the coefficient of variation (CV) of *effective* process time is inflated by equipment downtime and setups, rework and scrap and a host of other factors. Interruptions that cause long but infrequent periods of downtime will increase the CV more than ones that cause short, frequent periods of downtime, as long as the variability remains somewhat constant.

4. *Variability propagates.* If the output of a workstation is highly variable, it's inevitable that downstream workstations receiving products will also be highly variable.
5. *Waiting time is frequently the largest component of cycle time.* Two factors contribute to long waiting times: high utilization levels and high levels of variability. It follows, then, that increasing the effective capacity and decreasing variability will both work to reduce cycle times.
6. *Limiting buffers reduces cycle time, sometimes at the cost of decreasing throughput.* Because limiting inventory between workstation is the equivalent of implementing a pull system, it is the primary reason why reducing variability is so critical in Just In Time (JIT) systems.
7. *Variability pooling reduces the effects of variability.* Pooling variability will dampen the effects of variability, because it is less likely that a single occurrence will dominate performance.

The inevitable conclusion is that variability degrades the performance of a manufacturing organization. So where does this variability come from? Before we attempt to identify and locate sources of variation, it is important to first understand the causes of variability. It is equally important to be able to quantify it, and we can do this by using standard measures from statistics to define variability classes.

Again, Hopp and Spearman report that there are three classes of processing time variability, as seen in Table 9.1.

TABLE 9.1

Three Classes of Processing Time Variability

Variability Class	Coefficient of Variation	Typical Situation
Low (LV)	$CVt < 0.75$	Process times without outages (e.g. downtime)
Moderate (MV)	$0.75 \leq CVt < 1.33$	Process times with short adjustments (e.g. setups)
High (HV)	$CVt \geq 1.33$	Process times with long outages (e.g. failures)

When we think about processing times, we have a tendency to consider only the actual time that the machine or operator spends on the job, actually working (i.e. not including failures or setups), and these times tend to be normally distributed. If, for example, the average process time was 20 minutes and the standard deviation was 6.3 minutes, then $CV_t = 6.3/20 = 0.315$ and would be considered a low variation (LV) process. Most LV processes follow a normal probability distribution. Suppose the mean processing time was 20 minutes, but the standard deviation was 30 minutes. The value for $CV_t = 30/20 = 1.5$. This process would be considered highly variable.

You may be wondering why we care whether a process is LV, medium variation (MV) or high variation (HV). Suppose, for example, that you have identified a constraint that is classed as an LV process with an average process time of 30 minutes and a standard deviation of 10 minutes. The calculated value of the coefficient of variation, $CV_t = 10/30 = 0.33$ and would be considered low LV. Suppose that the non-constraint operation feeding the constraint has an average processing time of one-half that of the constraint, 15 minutes, but its standard deviation was 30 minutes. The calculated value for $CV_t = 30/15 = 2.0$ and is considered an HV process. A value of 2.0 from Table 9.1 suggests that this process probably has long failure outages, which could starve the constraint! When developing your plan of attack for reducing variation, using the coefficient of variation suggests that you include non-constraint processes that feed the constraint operation if they are classified as HV.

Although Hopp and Spearman write that these sources are found in manufacturing environments, it is my belief that they apply to virtually any environment where a process produces either products or delivers services. Hopp and Spearman [1] present five of the most prevalent sources of variation in manufacturing environments, as they apply to processing times:

1. *Natural variability*: includes minor fluctuations in process time, due to differences in operators, machines and material, and in a sense, is a catch-all category, since it accounts for variability from sources that have not been explicitly called out (e.g. a piece of dust in the operator's eye). Because many of these unidentified sources of variability are operator-related, there is typically more natural variability in a manual process than in an automated one. Even in a fully automated machining operation the composition of the

material might differ, causing processing speed to vary slightly. In most systems, natural processing times are low variability, so CV_t is less than 0.75.

2. *Random outages*: Unscheduled downtime can greatly inflate both the mean and the coefficient of variation of process times. In fact, in many systems, this represents the single largest cause of variability. Hopp and Spearman refer to breakdowns as *preemptive outages* because they occur whether we want them to or not (e.g. they can occur right in the middle of a job). Power outages, operators being called away on emergencies and running out of consumables are other possible sources of preemptive outages.

   Hopp and Spearman refer to *non-preemptive outages* as stoppages that occur between rather than during jobs and represent downtime that occurs, but for which we have some control as to when. For example, when a tool begins to wear and needs to be replaced, we can wait until the current job is finished before we stop production. Other common examples of non-preemptive outages include changeovers, preventive maintenance, breaks, meetings and shift changes. So how can we use these thoughts?

   Suppose we are considering a decision of whether to replace a relatively fast machine requiring periodic setups with a slower, more flexible machine that does not require setups. Suppose the fast machine can produce an average of one part per hour, but requires a two-hour setup every four parts on average. The more flexible machine takes 1.5 hours to produce a part, but requires no setup. The *effective capacity* (EC) of the fast machine is:

$$EC = 4 \text{ parts}/6 \text{ hours} = 2/3 \text{ parts/hour}$$

   The effective process time is simply the reciprocal of the effective capacity, or 1.5 hours. Thus, both machines have an effective capacity of 1.5 hours. Traditional capacity analysis would consider only mean capacity and might conclude that both machines are equivalent. Traditional capacity analysis would not recommend one over the other, but if we consider the impact on variability, then the flexible machine, requiring no setup, would be our choice (and that of Hopp and Spearman). Replacing the faster machine with the more flexible machine would serve to reduce the process time CV and therefore make the line more efficient and effective.

This, of course, assumes that both machines have equivalent natural variability.

3. *Setups*: The amount of time a job spends waiting for the station to be set up for production. Setups are like changeovers, in that they contain internal and external activities. Internal activities are those that must be done while the equipment is shut down, while external activities can be completed while the equipment is still running. The key to reducing setup time is to turn as many internal activities into external activities, thus reducing waiting time.
4. *Operator availability*: The amount of time a job spends waiting for an operator to be available to occupy the workstation and begin to produce product. The best way to reduce this type of time delay is to create a flexible work force. Having to wait for a specialist operator is no longer acceptable. Companies today must cross-train operators so that if one is called away, or is absent, another can step in and perform his or her tasks. This is especially critical in the constraint operation.
5. *Recycle*: Just like breakdowns and setups, rework is a major source of variability in manufacturing processes. If we think of the additional processing time spent "getting the job right" as an outage, it's easy to see that rework is completely analogous to setups, because both rob the process of capacity and contribute greatly to the variability associated with processing times. Rework implies variability, which in turn causes more congestion, work-in-process (WIP) and cycle time.

One of the keys to understanding the impact of variability is that variability at one station can affect the behavior of other stations in the process, by means of another type of variability referred to as *flow variability*. Hopp and Spearman explain that flow refers to the transfer of jobs or parts from one station to another, and if an upstream workstation has highly variable process times, the flow it feeds to downstream workstations will also be highly variable. In other words, variability propagates!

The concepts of processing time variability and flow variability are important considerations, as we attempt to characterize the effects of variability in production lines. But, it's important to understand that the actual processing time (including setups, downtime, etc.) typically accounts for only about 5 percent to 10 percent of the total cycle time in a manufacturing plant (Hopp and Spearman). The vast majority of the extra time is spent waiting for various resources (e.g. workstations, transporting, storage, operators, incoming parts, materials and supplies, etc.). Hopp and Spearman refer to the science of waiting as *queuing theory*, or the theory of waiting in lines. Since jobs effectively "stand in lines" waiting to be processed, moved, etc., it is important to understand and analyze why queuing exists in your process. Doesn't it make sense that, if waiting accounts for the vast majority of time a product spends in the system, then one of the keys to throughput improvement is to identify and understand why waiting exists in your process?

Queuing systems combine the impact of the arrival of parts from other processes and received parts from outside suppliers, the production of the parts and the inventory or queue waiting to be processed. Hopp and Spearman go into much depth on this subject, and I suggest you read their work, but the important thing to remember is this. Since limiting interstation buffers is logically equivalent to installing a kanban, this property is a key reason that variability reduction (via production smoothing, improved layout and flow control, total preventive maintenance and enhanced quality assurance) is critical to reducing variability, especially in the constraint.

Hopp and Spearman [1] point out that another, and more subtle, way to deal with congestion effects is by combining multiple sources of variability, known as *variability pooling*. An everyday example of this concept is found in routine financial planning. Virtually all financial advisers recommend investing in a diversified portfolio of financial investments. The reason, of course, is to hedge against risk and uncertainty. It is highly unlikely that a wide spectrum of investments will perform extremely poorly at the same time. At the same time, it is unlikely that they will perform extremely well at the same time. Hence, we expect less variable returns from a diversified portfolio than from a single asset.

Hopp and Spearman [1] go on to discuss how variability pooling affects batch processing, safety stock aggregation and queue sharing, but the important point to take away is this. Pooling variability tends to reduce the overall variability, just like a diversified portfolio reduces the risk of up and down swings in your earnings. The implications are that safety stocks can be reduced (resulting in less holding costs), or that cycle times at multiple-machine process centers can be reduced simply by sharing a single queue.

There are two basic but fundamental laws of factory physics relevant to variability provided to us by Hopp and Spearman.

1. **Law (Variability)**: Increasing variability *always degrades* the performance of a production system. This is an extremely powerful concept, since it implies that variability in any form will harm some measure of performance. Consequently, variability reduction is absolutely essential to improving the performance of a system. This is where Six Sigma comes into play.
2. **Law (Variability Buffering)**: Variability in a production system will be *buffered* by some combination of
   a. Inventory
   b. Capacity
   c. Time.

This law is an important extension of the variability law, because it specifies the three ways in which variability impacts a manufacturing process and the choices we have in terms of buffering for it.

The primary focus of this section was to explain the effect of processing time variability (PTV) on the performance of production lines. The primary points, conclusions or principles are provided, once again, by Hopp and Spearman:

1. *Variability always degrades performance.* As variability of any kind is increased, either inventory will increase or lead times will increase or throughput will decrease or a combination of the three. Because of the influence of variability, all improvement initiatives must include variability reduction. As presented in earlier chapters in [1] *The Ultimate Improvement Cycle*, there are important steps that focus on variability reduction.
2. *Variability buffering is a fact of manufacturing life.* If you can't reduce variability then you must buffer it, or you will experience extended cycle times, increased levels of inventory, wasted capacity, reduced throughput and longer lead times, all of which result in declining revenues, missed delivery dates and poor customer service.
3. *Flexible buffers are more effective than fixed buffers.* By having capacity, inventory or time available as buffering devices, you now have a flexible combination of the three to reduce the total amount of buffering needed in a given system. Examples of each type of buffer are in Table 9.2.
4. *Material is conserved.* Whatever flows into a workstation must flow out as acceptable product, rework or scrap. It's either good, bad or reworkable product, but obviously we prefer only good product.

TABLE 9.2

Buffer Types

Flexible Buffer Type	Buffer Example
Flexible capacity	Cross-trained workforce—by moving flexible workers to operations that need the capacity, flexible workers can cover the same workload with less total capacity than would be required if workers were fixed to specific tasks.
Flexible inventory	Generic WIP held in a system with late product customization. That is, having a product platform that results in potentially different end products.
Flexible time	The practice of quoting variable lead times to customers depending on the current backlog of work (i.e. the larger the backlog, the longer the quote). A given level of customer service can be achieved with shorter average lead time if variable lead times are quoted individually to customers instead of uniform fixed lead time quoted in advance. This is possible if you significantly reduce your cycle time to the point that the competition can't match it.

5. *Releases are always less than capacity in the long run.* Although the intent may be to run a process at 100 percent of capacity, when true capacity, including overtime, outsourcing, etc., is considered, this really will never occur. It is always better to plan to reduce release rates before the system "blows up" simply because they will have to be reduced as a result of the system "blowing up" anyway. This is why Drum Buffer Rope works so well … it prevents WIP explosions.
6. *Variability early in a line is more disruptive than variability late in a line.* Higher front-end process variability of a line using a push system will propagate downstream and cause queuing later on in the process. By contrast, stations with high process variability toward the end of the process will affect only those stations. Remember, variability propagates, so the further upstream the variability occurs, the more disruptive its effects will be.
7. *Cycle time increases nonlinearity in utilization and efficiency.* As utilization approaches 100 percent, long-term WIP and cycle time will approach infinity. Companies that attempt to drive the total process utilization and/or efficiency higher and higher will clearly have problems with excessive WIP, long cycle times, missed delivery dates and poor customer satisfaction … plus huge quality problems. It's why I loathe efficiency and utilization as a performance metric in any place other than the system constraint.

8. *Process batch sizes affect capacity.* Increasing batch sizes increases capacity and thereby reduces queuing, while increasing batch size also increases wait-to-batch and wait-in-batch times. Because of this, the first focus in serial batching situations should be on setup time reduction, enabling the use of small, efficient batch sizes. If setup times cannot be reduced, cycle time may well be minimized at a batch size greater than one. In addition, the most efficient batch size in a parallel process may be in between one and the maximum number that will fit into the process. The bottom line here is, the whole idea of economic batch quantity is riddled with wrong assumptions … but don't try and convince your local Cost Accounting group of this because they'll accuse you of a sacrilege.
9. *Cycle times increase proportionally with transfer batch size.* Because waiting to batch and un-batch is typically one of the largest sources of cycle time length, reducing transfer batch sizes is one of the simplest and easiest ways to reduce cycle times. Instead of waiting for the full batch to be produced and moved to the next process step, move product periodically to the next step so it can be worked. But then again, I don't believe in large batches at all. To me, the key is to simply reduce your process batch size and focus efforts on reducing changeover times through SMED.
10. *Matching can be an important source of delay in assembly systems.* Lack of synchronization caused by variability, poor scheduling or poor shop floor control will always cause significant build-up of WIP, resulting in component assembly delays.

MORE ON VARIABILITY

We're all familiar with the positive effects of implementing cellular manufacturing in our workplaces such as the improved flow through the process, the overall cycle time improvement, throughput gains as well as others. But there is one other positive effect that can result from cellular manufacturing that isn't discussed much—the potential positive impact it can have on variation. When multiple machines performing the same function are used to produce identical products, there are potentially multiple paths that parts can take from beginning to end. There are, therefore, potential multiple *paths of variation.*

These multiple paths of variation can significantly increase the overall variability of the process.

Even with focused reductions in variation, real improvement might not be achieved because of the number of paths of variation that exist within a process. Paths of variation, in this context, are simply the number of potential opportunities for variation to occur within a process because of potential multiple machines processing the parts. And the paths of variation of a process are increased by the number of individual process steps and/or the complexity of the steps (i.e. number of sub-processes within a process).

The answer to reducing the effects of paths of variation should lie in the process and product design stage of manufacturing processes. That is, processes should/must be designed with reduced complexity, and products should/must be designed that are more robust. The payback for reducing the number of paths of variation is an overall reduction in the amount of process variation and ultimately more consistent and robust products. Let's look at a real case study.

Several years ago, I consulted for a French pinion manufacturer located in southern France. When I arrived at this company, it was very clear that it was being run according to a mass production mindset. There were multiple, very large containers of various sized pinions stacked everywhere. The process for making one particular size and shape pinion was a series of integrated steps from beginning to end as depicted in Figure 9.1. The company received metal blanks from an outside supplier, which were fabricated in the general shape of the final product and then passed through a series of turning, drilling, hobbing, etc. process steps to finally achieve the finished product. The process for this particular pinion was automated with two basic process paths, one on each side of this piece of equipment. There was an automated gating operation that directed each pinion to the next available process step as it traversed the entire process, which consisted of 14 steps.

It was not unusual for a pinion to start its path on one side of the machine and move to the other side and back again, which meant that the pinion being produced was free to move from side to side in random fashion. Because of this configuration, the number of possible combinations of machines used to make the pinion, or paths of variation, was very high. Let's take a look now at the number of paths of variation that existed on this machine as seen in Figure 9.1.

The first step in the process for making this style pinion was an exterior turning operation with two turning machines available to perform this

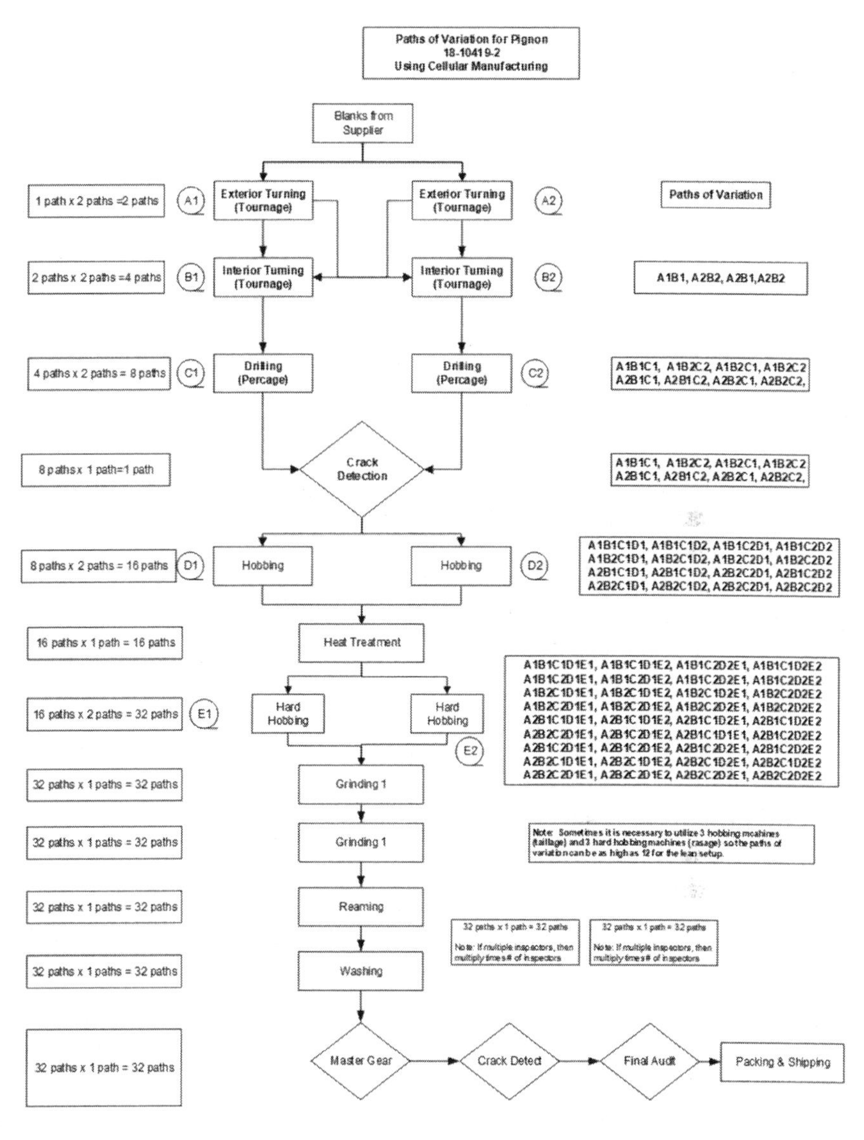

FIGURE 9.1
Possible Paths for a Pinion.

function (labeled A1 on one side of the machine and A2 on the other side of the machine as shown in Figure 9.1). This purpose of this first step, like the others to follow, was to shave metal off of the blank to ultimately achieve its final shape and critical dimensions. The next step in the process is referred to as interior turning, and again, there are two interior turning machines labeled B1 and B2, one on each side. In the third step, there were two possible choices for drilling, C1 and C2. After the pinion was automatically inspected for cracks (with one, common automated gage), it then progressed to one of two hobbing machines, D1 and D2. The parts were then collected in storage bins and sent as large batches to an outside vendor for heat treatment. Upon return from heat treatment, the pinions then proceeded to hard-hobbing, E1 and E2, and then on through the remainder of the process as indicated in Figure 9.1.

The boxes to the right in Figure 9.1 represent the possible paths that the pinion could take as it makes its way through the process. For example, for the first two process steps, there are four possible paths, A1B1, A1B2, A2B1, and A2B2. In Figure 9.1 the possible paths of variation are listed to the right, and as you can see, as the part continues on, the possible paths continue on until all 32 potential paths are seen. Do you think that the pinions produced through these multiple paths will be the same dimensions, or will you have multiple distributions? What if we were able to reduce the number of paths of variation from 32 down to two, do you believe the overall variation would be less and how many distributions would you have now? Or another way of saying this, do you believe the part-to-part consistency would be greater? Let's check it out.

In Figure 9.2, we created what we'll call a "virtual cell," meaning that we limited the paths of travel that an individual pinion can take by removing the possibility for a part to traverse back and forth, from side to side, in this machine configuration. In simple terms, the part either went down side 1 or side 2, rather than allowing the gating operation to select the path. In Figure 9.2, you can see that pinions passing through turning machine A1 are only permitted to proceed to turning machine B1. Those that pass through turning machine A2 are only permitted to proceed to turning machine B2. In doing so, the number of paths of variation for the first two process steps was reduced from four to two.

Continuing, the parts that were turned on A1 and B1 can only pass through hobbing machine E1, while those produced on A2 and B2 can only be processed on hobbing machine E2. To this point, the total paths of variation remain at two instead of the original number of paths of 16. The part continues to hard-hobbing where there are, once again, two machines available. The parts produced on A1B1E1 can only proceed to the G1 hard-hobbing machine, while those produced on A2B2E2 can only be processed on hard-hobbing machine G2. We also instructed the heat-treater to maintain batch integrity and not mix the batches. So at this point, because we specified and limited the pinion paths, the total paths of

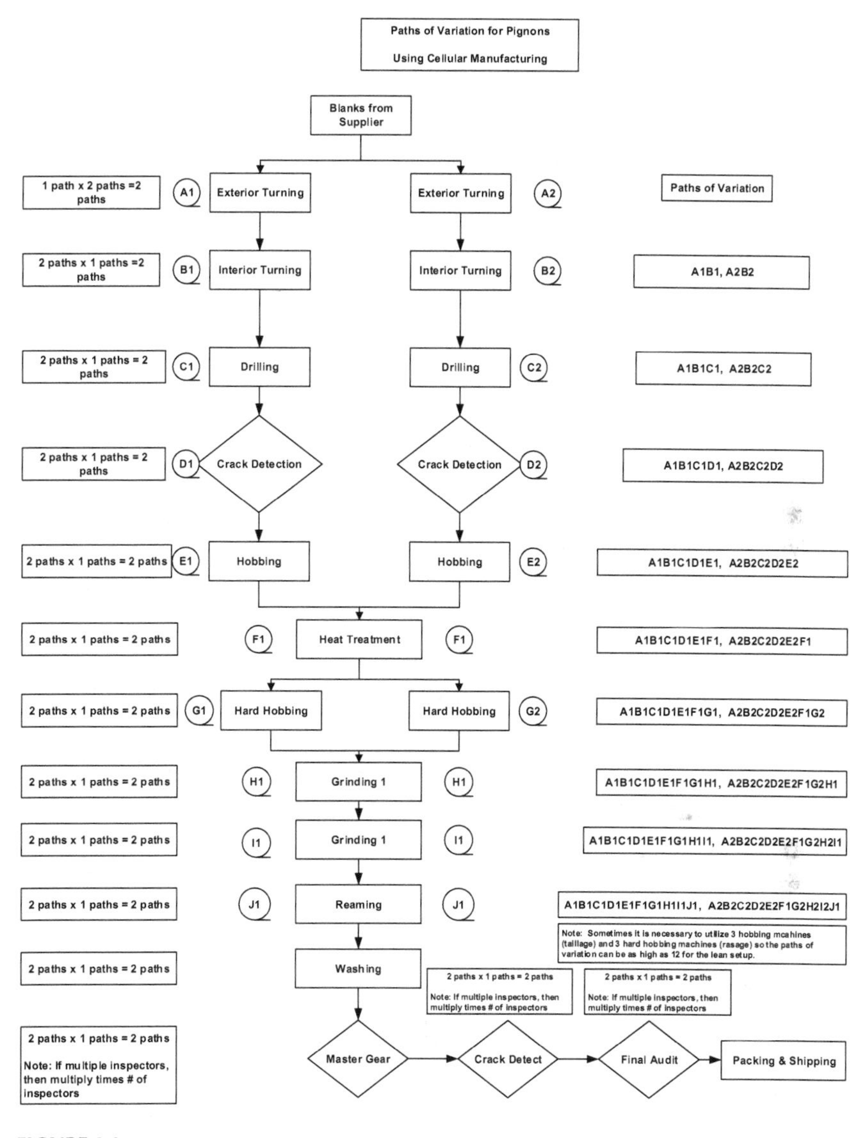

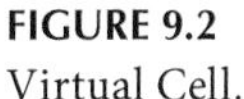

FIGURE 9.2
Virtual Cell.

variation decreased from 32 to only two! So, what do you think happened to variation when we created our virtual cell?

The key response variables for this process were five individual diameters measured along the surface of the pinion. As a result of limiting the number of potential paths of variation, the standard deviation for the various diameters was reduced by approximately 50 percent on each of the diameters!

But, even though we were very successful in reducing variation, there was a problem associated with making this change, sort of an unintended consequence, if you will. Remember in the original configuration, pinions could move to the next available machine (i.e. either side) as they proceeded along the flow of the process. With the new configuration, they could no longer do this. Prior to this change, when there was downtime on one side of the machine, the automation and/or operator simply diverted the pinion to the same machine on the other side, so as to keep the parts moving. With the new configuration, when a machine in the cell went down unexpectedly, the parts now had to wait until the machine was repaired.

The immediate short-term result of this change was a significant reduction in throughput of pinions, because of unplanned downtime. However, in the longer term, it forced the company to develop and implement a preventive maintenance system, which eventually reduced the unplanned downtime to nearly zero. When this happened, the new throughput surpassed the original throughput, and the variation was reduced by 50 percent! In addition, the scrap levels for this process were reduced by 40 percent! And the great part was, not a single dollar—or should I say Euro—was spent in doing this, yet the payback was huge. So, as you're studying your process for variation reduction, keep the concept of paths of variation foremost in your mind, because it can make a huge difference in some circumstances.

REFERENCE

1. Hopp, Wallace J. and Spearman, Mark L., *Factory Physics*, McGraw-Hill, New York, 2001.

10

Performance Metrics

INTRODUCTION

I've been asked many times which of Goldratt's 5 Focusing Steps is the most difficult for companies to embrace. From my perspective and experience it's Step 3, which deals with subordinating everything to the constraint. One of the major reasons for this is based in Cost Accounting (CA), or even more specifically, the performance metrics that CA embraces and mandates. In particular, the performance metric *Operator Efficiency* in places other than the constraint is, in my mind, the most difficult mindset to overcome for many companies.

Step 3 of Goldratt's 5 Focusing Steps tells us to subordinate non-constraints to the constraint operation. So, in real terms, what does that really mean? If we were to look in a dictionary for the definitions of *subordinate*, we would probably find something like "*Belonging to a lower or inferior class or rank; secondary or subject to the authority or control of another.*" The fact that a subordinate is secondary and subject to the authority or control of another, really says it all. Literally, the constraint operation should dictate everything that goes on in a system and process. Of course, you want the constraint working at the rate required to satisfy customer orders, but the non-constraint operations should work only at the pace that the constraint does, neither faster nor slower.

Upstream non-constraints have the responsibility of never letting the constraint operation run out of material, because the throughput of the system would stop and so would new incoming revenue. But by the same token, the non-constraint operations should not work faster than the constraint, simply because running faster would not result in more system throughput. In fact, running faster would end up costing the operation more because of the added carrying costs of excess inventory, extended cycle time and, ultimately, a drop in on-time deliveries.

In reality, subordinating non-constraint operations to the constraint operation involves taking a back seat to the constraint. For example, if the constraint operation is experiencing equipment downtime, maintenance must drop everything else, move to the constraint and get the equipment back on line. If there is a quality problem on the constraint operation, the focus must be on correcting the problem, even at the expense of a non-constraint quality problem. Because the non-constraint operations have "sprint capacity," in the event of downtime they should be able to shut down and wait for resources, without negatively impacting throughput at the constraint.

Subordination is not limited to typical manufacturing functions either. For example, engineering must always be available to move to a constraint-based problem, at the expense of non-constraints. Sales must be constantly aware of and pursuing new orders that will keep the constraint operation busy, but not beyond its capacity. In effect, all resources, which include all functional groups and non-constraint operations, must be available to actively support the needs of the constraint operations, without exception. Remember, a constraint operation should never be left idle, because lost production can *never be recovered*.

One question I have often been asked is, "Why is it so important to subordinate the non-constraint resources to the constraints?" At the risk of being redundant and belaboring the point, *because the system constraint determines the overall system Throughput and the ultimate profitability of the organization*. To quote Debra Smith [1], in her great book, *The Measurement Nightmare*, "The ability to subordinate will define a company's ability to succeed with the Theory of Constraints. Exploitation of the constraint is dependent on effective subordination."

Smith further states, "Disruptions cause waste and accumulation of delays. Delays and waste cause costs to go up and Throughput to go down." Any improvement initiative that utilizes all or parts of the Theory of Constraints (TOC) must be locked into and committed to the concept of subordination. Without embracing subordination, at some point in time the constraint operation will be starved, Throughput will decrease, and profits will be less than optimal. And remember, lost Throughput is lost forever and can never be regained.

Improvement efforts that focus on areas other than the constraint operations are, for the most part, generally fruitless. That is, improving the output of a non-constraint operation not only increases carrying cost of inventory, but also lengthens the product cycle time and causes deterioration of on-time delivery rates. Although the other parts of the system could produce more, there simply is no point in doing so. Subordination changes the way an organization carries out its business and redefines the objectives in every part of the global system.

As I mentioned earlier, material should be released according to the needs of the constraint only and according to firm orders. Maintenance priorities will be set according to the needs of the system constraint. Manufacturing engineering will be at the beck and call of the constraint. Every group and employee in the organization must recognize that the key to short- and long-term profitability, and hence realization of the organization's goal, is the recognition and belief that the system constraint is the *first priority*.

Now that you have an idea of what subordination is, and why you should subordinate the non-constraints to the constraint, the question then becomes one of how you do it. But before I answer this question, there are some basic guidelines—or maybe a better word is *truths*—to discuss. And for your typical American company, these guidelines or truths may be a difficult pill to swallow, because the leadership and management of many companies have been taught, rewarded and disciplined using these basic beliefs for many years.

- If your company is currently using manpower efficiency or equipment utilization as its primary performance metric, be prepared to abandon or at least scale back its relative importance significantly in your non-constraint operations. Measuring the efficiency of a non-constraint may result in local improvements, but it will not ensure that the overall performance of the organization will improve. In fact, focusing on improvements in non-constraint efficiencies will actually cause the performance of the total organization to become worse, not better. Just remember that system improvement is not the sum of local improvements.
- Because you will now be optimizing and making decisions about the organization from a system perspective, rather than optimizing locally like so many companies do, be prepared to have some of your non-constraint operators and machines sitting idle at different periods of time for the good of the total organization. The important point to remember here is, contrary to some beliefs, an idle resource is not a significant source of waste. I realize this is contrary to what Lean and Cost Accounting teaches us, but the fact is, 100 percent utilization of non-constraints is not a good thing.

- The reality is that constraints do exist, and if you do not manage them, they will manage you. If you do not manage them, your organization will be in a constant fire-fighting mode. Managing constraints is one of the keys to organizational and operational success.
- Focusing resources on a non-constraint process will not result in a maximization of a company's return on investment. Focusing resources on the system constraint to improve performance will improve the performance of the total system and will maximize the return on investment.
- Be prepared to find many nonphysical constraints. In fact, most of the constraints you will identify will probably be rooted in outdated policies and beliefs that probably were effective at one point in time, but have outlived their usefulness and are now hurting the organization. These policies are usually traced to things like staffing decisions; how products and materials were scheduled, purchased and supplied; product pricing policies; how performance is measured; and how people in the organization should be managed. All these type constraints usually are based on flawed and outdated assumptions and many times they are observed as unwritten rules that have been passed down from previous leadership. But beware, nonphysical constraints are not necessarily easy to correct. In fact, most times they are much more difficult to correct, simply because they have become part of the DNA of the organization.

But fear not, subordinating non-constraints to constraints may not be as difficult as you might think or as difficult as some writers have suggested. Dettmer, in his classic book *Breaking the Constraints to World Class Performance* [2], presents what he calls a TOC decision matrix. In this matrix, Dettmer converts three traditional global measures (i.e. net profit, return on investment and cash flow) into TOC terms and local measures of optimization. He then lists a series of questions, which require a yes or no answer. Dettmer concludes that if the answer to the question is yes, do it. If the answer is no, do not do it ... pretty straightforward. These questions can be used to evaluate local decisions and answer the question of how to subordinate non-constraints to the constraint operation. Dettmer's questions that require only a yes or no answer are:

- Will it increase sales?
- Will it speed up deliveries to clients?

- Will it reduce backlogs?
- Will it reduce your need for production materials?
- Will it shorten production time?
- Will it reduce fixed expenses?
- Will it shorten the time between product or service delivery and time of payment?
- Will it increase the volume of revenue received in the same time period?
- Will it shorten the time between receipt of order and delivery to the customer?
- Will it free excess capacity?
- Will it make better use of the constraint?
- Will you need fewer materials on hand?
- Will you need less equipment?

When considering any kind of improvement, or in this case, considering an action aimed at subordinating a non-constraint to a constraint operation, just ask and answer these questions, and you will make the right decision.

EFFICIENCY, PRODUCTIVITY AND UTILIZATION (EPU)©

Many businesses, seemingly across all industries, are prone to develop and use some type of metrics in their decision-making process. Many of these organizations are focused on Efficiency, Productivity and Utilization, or EPU. Using these metrics as guidelines, many organizations try to forecast their business activity and make a judgment call concerning their current status. And in many respects, this is probably not all bad. However, what is bad is the seemingly common nonsensical way these metrics are used.

Many business leaders understand that, to make good business decisions, they must have good data on which to base their decisions. If the data is incorrect or interpreted incorrectly, it is highly probable that bad decisions will be the outcome. And if bad decisions are implemented, it can spell the death of an otherwise good company.

If good data is required to make good decisions, then collecting, interpreting and using the data would seem to be of paramount importance. Useful data collection is a way of using past performance to help predict future performance. Accurate data can provide the user with the ability to make the necessary mid-course corrections and get the organization back on track and headed in the right direction. This brings up another important point. It's imperative to know where you are going to set a course of actions on how to get there. If you don't know where you are going, then it doesn't matter what actions you take to get there. Unfortunately, this seems to be the way that many organizations make crucial decisions! Many organizations make decisions not based on what they need or understand, but instead based on what everyone else is doing. This decision-making process is sometimes referred to as "bench marking." This type of decision thinking only works if you can validate the assumption that what the competitor is doing is correct! It is possible that what the competitor is doing might make sense for them, but not necessarily for you. So, wishfully following what the competitor is doing in hopes of having the same effect for your organization is fantasy leadership.

Many organizations use some kind of metric on a daily basis, albeit in a roundabout way. Consider the instrumentation in your car. There are measuring devices (gages and read-outs) designed to keep the driver informed as to the operational stability of the vehicle they are driving. The gas gage tracks the consumption of raw material inventory required to keep the system operational. The tachometer tracks the speed at which the system is working (RPM—revolutions per minute), and the speedometer monitors the speed of the system through time. The temperature gage, oil pressure gage and battery charging system all provide vital data about the status of the system, but only if you understand how to interpret the data and react to it. If you analyze and interpret the data incorrectly, the system could operate suboptimally, or worst case the system could fail. That is not what we want to have happen. These same measurement principles hold true for analyzing production systems and business data within an organization. So, what is important to measure, and why is it important?

EFFICIENCY

Somewhere along the way, efficiency became "king" of the metrics mountain. Many organizations are measuring efficiency, not because it was really important to them, but rather because somebody else was measuring efficiency (bench marking)—the assumption being that if the competitor is doing it, then we should be doing it also. Question: "Is that a valid assumption?"

Efficiency is a metric used by many industries, and it is a metric that is used incorrectly most of the time. When used incorrectly, efficiency will give the false impression of "If we look good, then we are doing good." There is hardly a day that goes by that you don't read about efficiency in the paper, or hear it used on the news. The new battle cry is, "We must become more efficient at what we do." Or, "We must improve our efficiency to stay competitive." There is a downside to these mottos if, in fact, you measure efficiency incorrectly.

The concept of "efficiency" is often confused with the term "effectiveness." Many believe that a high efficiency is synonymous with being highly effective. Not true! Efficiency is measurable and therefore quantitative. Effectiveness is non-quantitative and therefore a rather vague concept associated mostly with achieving a goal or objective.

Efficiency also has many models for application including Physics, Economics and other sciences. It is expressed in terms of a ratio, i.e. the ratio in terms of something *produced* (units) divided by the resources *consumed* to produce it (hours), or $r = P/C$. However, there are some concerns with trying to use the efficiency model in a production setting. The most obvious is that the efficiency model doesn't fit well within a typical production system. The best measure of the production system is the productivity measure. In essence, the productivity measure tells you the efficiency of the system—but more on that later.

The mathematical limitation of efficiency is that it can never exceed 100 percent, and yet there are companies who proclaim much higher efficiency metrics than 100 percent! How do they do that? One simple trick they employ is to measure efficiency as a ratio of standard hours given divided by actual hours used. As an example, suppose for a period of work time there were 1000 standard hours issued to do the work, but the actual time to do the work was tracked at 500 hours, then $1000/500 = 200$ percent efficiency. The first hint that this is incorrect comes from the definition of efficiency—there is no variable of what is actually *produced*! Measuring efficiency this way only gives you a ratio of the hours given (standard hours) compared to hours used (actual hours). If efficiency goes up (which it probably will), it translates to the standard hours being incorrect (which they probably are.) In fact, when using this method (and many do) of calculating efficiency, it is very probable that ALL the standard hours could be consumed and not a single unit of product produced! The metric would tell you that the system is operating at 100 percent efficiency, and yet not a single unit was produced. What the measure really says is this: "*The more you improve the further away from the goal you get!*" Is this an accurate measure of the system performance?

In this case, the system looks really good (the efficiency), but the system isn't performing well at all (the output.) So, how useful is a metric that paints this picture? The efficiency metric says you are doing fine, and yet reality says you are missing the goal/objective! Suppose this was the data output from your system—what would you do?

For obvious reasons, the efficiency metric is not reliable, and certainly not an accurate measure to get a clear picture of what is going on. Even when efficiency is calculated correctly, it can still have a devastating effect on a system. As an example, suppose in your organization the metric was to maintain high efficiency levels *all* the time. In this scenario, efficiency can be literally interpreted to mean "*keep everyone busy all the time.*" The assumption with this thinking is that "*busy people*" equate to high efficiency—which is true! But keeping people busy all the time also prompts an organization to buy and release more and more raw materials to achieve its goal. Buying and releasing more raw materials only serves to increase the work-in-process (WIP) in the system. Higher WIP levels will also have a negative effect on on-time delivery (OTD) and will cause it to drop as the WIP levels go higher.

A system can actually become so polluted with WIP that it might produce nothing at all. So, you achieved a very high efficiency, but at what expense? Understanding these different scenarios begs the question, "Does high efficiency also equal higher levels of productivity?" Efficiency when used this way could prevent good decision making for production systems. The efficiency metric would better serve its user for calculating the gas mileage of a car, or for the efficiency of a gas furnace, but not so much in a production system. So, if efficiency might not be the best metric for a system, then what is? Let's take a look at the merits of the productivity metric and see how it might apply.

PRODUCTIVITY

Productivity is another one of those metrics that has multiple uses and multiple ways of being calculated, depending on how it is being used. You can use it for economic productivity, labor productivity, total factor productivity, service sector productivity and several other forms of measurement. But for our discussion, we want to look at the productivity metric as it applies to a production system. We're looking for a metric to measure the system and answer the question, "Did we improve?" For a production system application, we can define productivity as the ratio between output and input, or simply

$$\text{Productivity} = \text{Output quantity}/\text{Input Quantity}\left(p = O/I\right)$$

Mathematically, the answer is a ratio and is calculated by determining the number of units completed (output), divided by the number of units input (usually hours). As an example, suppose you had completed 750 units of work and used 1000 hours completing the work the equation would look like this:

$$0.75 = 750 \text{ units}/1000 \text{ hours}$$

In essence, for every hour worked you completed 75 percent (0.75) of one unit of work. In this instance, the productivity measure is also the efficiency measure of the system. Now when the system is measured, we can answer the question "Did we improve?" The best means to improving productivity is to increase the numbers of units produced while the measured time stays the same or decreases. Or you could decrease the amount of time required to make the same number or more units.

A closer look reveals that productivity can actually be sub-divided into two separate processes. The first is the *production process*, and the second is the *monetary process*. The production process is a measurable component of the system, but the monetary process is also measurable, if you do it correctly. In the Theory of Constraints measurements library, there is a formula for measuring productivity. The formula is simple and is expressed as follows:

$$\text{Productivity} = \text{Throughput}/\text{Operating Expense or } \left(P = T/\text{OE}\right)$$

For this to work, both the output and input need a common denominator, which in this case is dollars. The output can be measured in dollars by using the Throughput calculation. Throughput is defined as "*the rate at which inventory is converted to sales.*" The monetary calculation for Throughput is the selling price of the product minus the total variable cost (TVC). Total variable cost is defined as the cost of raw material, plus any sales commission (for each product sold), plus the shipping costs. In other words, you are looking for all the costs associated with a single product. However, the labor costs are not included in the TVC. The labor costs are part of the Operating Expense (OE) dollars. The difference between the Selling Price (SP) and the Total Variable Costs (TVC) is considered to be the Throughput (T). As an example, suppose we had a product that sold for $1.00 and the TVC was $0.40, then the Throughput (T) would be $0.60.

The Operating Expense can be expressed in dollars and is calculated as labor, overhead, gas, lights, benefits and all other expenses. The attractiveness of using this type of measure is that it includes *ALL* of your expenses, and not just the labor expense in the form of hours used. This measure provides a much cleaner picture of the productivity and not just a partial look. Also, by knowing the Throughput and Operating Expense numbers it becomes very easy to calculate the Net Profit (NP) during any given period. Net Profit is simply determined by subtracting the Operating Expense from the total Throughput number. In other words:

$$\text{NP} = (T - \text{OE}).$$

The important aspect of accurately determining the productivity metric is converting units to Throughput dollars and Operating Expense to dollars. The dollars component is the common denominator between units and hours. Once you have that information, the productivity measurement accuracy improves greatly. Let's move on to a discussion about Utilization.

UTILIZATION

Utilization is most often the metric that organizations should be using instead of erroneously using the efficiency metric. Unlike efficiency and productivity, which are ratio calculations based on input and output, the utilization metric is a proportional measure of resource time used divided by resource time available, simply stated as

$$\text{Utilization} = \text{Actual time used}/\text{Time available}$$

As an example, suppose we wanted to measure machine utilization for a specific machine in our system. Suppose this machine was available to work one full shift, which for our example would equal one shift of eight hours in duration. During this eight-hour shift there are 480 minutes available to work (8 hours × 60 minutes = 480 minutes). Of those 480 minutes available, let's say our machine was busy making parts for 395 minutes; then,

$$395(\text{Time used})/480(\text{time available}) = 82.29\% \text{ Utilization.}$$

In many organizations, lower utilization percentages (if they measure them at all) are considered as inefficient and should be improved to operate as close to 100 percent as possible. It is possible that lower utilization numbers can be attributed to lengthy setup time or periods when an operator is not available (reasons can be, and are, varied), or simply periods of time when the machine has nothing to do. For many organizations this is an unacceptable situation, and they take actions to correct it. So to improve the utilization (efficiency) of the machine (resource), more work is released to keep the machine active. Most organizations take these actions not because they *must*, but solely because they *can*. The only reason this action is taken is to preserve the internal utilization/efficiency measure. However, there is an important lesson to be learned here, and that is, *activation of a resource does not necessarily equal utilization of the same resource*. In other words,

$$\text{Activation} \neq \text{Utilization}$$

Activation is using a machine because you can. *Utilization* is using a machine because you must! There is a vast difference between these two approaches. If you understand your system, then you understand that high utilization should only be considered at a constraint location and not implemented at non-constraint locations. It's not important that utilization numbers be at or near the 100 percent mark for all the resources in the system. Those actions would be counterproductive to system output. What is important is that utilization is implemented and monitored at the constraint location of the system. Subordination should be the implemented rule for all non-constraint locations. In other words, let the non-constraint work at a level necessary to support the needs of the constraint. No more—no less.

SUMMARY

Are these metrics useful or evil? The answer is—it depends. It depends on if you calculate and use them correctly. If these metrics are used to define a baseline measure, the metric can help you answer the question, "Did we improve?" Used incorrectly, these metrics will result in poor decision making.

In a production system, the productivity metric is probably the most important, if used correctly. It is possible to combine the sub-elements of productivity and define the common denominator of dollars for both the *production process* and the *monetary process*. Using productivity in this fashion will give you the clearest and most accurate assessment of your system.

The utilization measure is best used to monitor the constraint in a system, but only the constraint and not the non-constraints. It's a good metric, but not when it is applied everywhere. Implementing utilization at the constraint will help you *focus* on the most important location in your system. By monitoring this location, you can determine how much *leverage* you have in the system to meet the growing load demand.

Efficiency is by far the worst metric in a production system. It's just not a good fit. It's like trying to put a size nine shoe on a size ten foot. It requires a lot of manipulation, and it doesn't always feel right. It usually creates the opposite behavior from what you are really trying to achieve in the system.

Performance metrics are intended to serve some very important functions:

1. Performance measure should be designed and selected based on the behaviors you want exhibited in your organization.
2. Performance measures should reinforce and support the goals and objectives of the organization or company.
3. The measures should be able to assess, evaluate and provide feedback as to the status of people, departments, products and the total company.
4. Performance measures should be objective, precisely defined and quantifiable.
5. The measures should be well within the control of the people and/or departments being measured and not some abstract number.

6. Performance metrics must be understood and utilized by the organization as a whole, and they must positively impact the system and not individual parts of it.

We will discuss each of these functions in more depth shortly, but first I want to clarify this last function. This last function, especially that they be understood and impactful to the system and not parts of it, is very important. If people don't understand the metric, they simply won't understand the behaviors that are required to move it in the right direction. I also believe that companies should develop a hierarchy of sub-metrics so that even people at the lowest rung in the organizational ladder will understand how their behavior drives the metric in the correct direction. Let's look at an example.

Suppose your company has selected the performance metric efficiency and you are a production manager. You're told that your performance appraisal is based upon how high your production unit's efficiency is. If this was your mandate, how would you make this happen, or what behavior would you exhibit to reach your highest performance? And remember, your personal appraisal is based on how high your unit's efficiency is.

If it were me in this position, I'd probably tell my boss that this isn't a good metric to measure me. But most people would look at the metric and say to themselves, "If I want higher unit efficiencies, then I must run all of my process steps as fast as I can." So, what would that do to the unit? Well, for one, it would drive efficiencies higher and put me in a position to get a good appraisal. But what would it do to the organization? Think about it. If I run all of the process steps as fast as I could, what would be the organizational impact? Would we produce and ship more product? Would we spend less money? What?

Quite simply, if you run all process steps as fast as you can, the net effect would be as follows:

1. Your unit's efficiency metric would move to its highest level, good for your personal appraisal.
2. Work-in-process inventory would grow larger and larger.
3. Because WIP grows larger, cycle times become extended.
4. Because cycle times become extended, on-time deliveries decrease proportionally to the level of WIP.
5. Because on-time deliveries have decreased, customer satisfaction levels fall.
6. Because customer satisfaction levels fall, sales will decrease.

Need we go any further? Yes my friends, selecting the "right" performance metrics is critical to your long-term survival, so reason them out before you select them, and above all else, make sure the metric has the total organization in mind.

As stated earlier, selection of the right performance metrics is absolutely critical to the success of any organization, no matter whether they produce products or deliver a service. There are three key objectives of performance metrics as follows:

1. *First and foremost, performance metrics should stimulate the right behaviors.* Sounds simple enough, but unless the desired behaviors are well thought out, it is very easy to go astray. For example, if your organization produces products and uses the performance metric *operator efficiency*, ask yourself what behaviors should you expect to see? Translated, efficiency deals with minimizing waste and maximizing the capabilities of your human resources. From a Cost Accounting perspective, it is believed that improving operator efficiency has a direct impact on the profit of a company. So, if we increase operator efficiency, then we should see a corresponding increase in profits … right?

   The problem with operator efficiency as a performance metric is that it doesn't consider the impact on the total system. The behavior we typically see when using efficiency as a metric is this. Because all of the process steps are encouraged to "run to their maximum capacity," the total system is flooded with excess work-in-process inventory which extends cycle times which negatively impacts on-time delivery. So, do you think efficiency stimulates the right behavior? The answer is no, it doesn't. But having said that, what if we only measured the efficiency of the system constraint? What would happen then? We would indeed maximize the Throughput of the process because Throughput is controlled by the system constraint. What about the non-constraint process steps? Because we don't want them to outpace the constraint, they must effectively "slow down" so as not to fill the system with excess WIP. Their efficiencies would deteriorate, but the profit of the overall system would improve dramatically. This is directly in contrast to what traditional Cost Accounting teaches us. That is, as operator efficiency increases, profits rise accordingly.
2. *Performance metrics should reinforce and support the overall goals and objectives of the company.* If the goal of any for-profit company

is to make money now and in the future, then the selection of performance metrics must directly support and enhance this goal.

3. *The measures should be able to asses, evaluate and provide feedback as to the status of people, departments, products and the total company.* The right behaviors of people and departments are critical to the achievement of the overall goal of the company, but often the metrics chosen encourage and stimulate the opposite behaviors ... just as I demonstrated earlier with operator efficiency. The fact is, efficiency drives local optimization rather than optimization of the total system.

When selecting performance metrics, there must be criteria for selection of the correct ones, right? Well, in fact, there are at least six key criteria to consider when selecting effective performance metrics. Let's look at these criteria and relate each one to operator efficiency.

- The metric must be objective, precisely defined, measurable and quantifiable. There can be no ambiguity at all with the people or departments being measured. For example, operator efficiency is objective, well defined, measurable and quantifiable, so it would seem to satisfy this first criteria ... right?
- The metric must be well within the control of the people or departments being measured. For example, when considering the metric operator efficiency, it clearly is within the control of the people or department being measured, so it would seem that it does satisfy this criteria ... right?
- The metric must be translatable to everyone within the organization. That is, each operator, supervisor, manager and engineer must understand how his or her actions impact the metric. For example, efficiency is definitely translatable to everyone, so again, efficiency passes this litmus test ... right?
- The metric must exist as a hierarchy so that every level of the organization knows precisely how their work is tied to and supports the goal and critical success factors of the company. For example, if one of the critical success factors was "Maximum Throughput," and efficiency was selected as one of the lower level metrics, what would happen? That is, if it takes five minutes to process a part in an individual workstation, then the maximum amount of time for the part to be finished should be no longer than five minutes. And

if you could somehow produce a part in four minutes, the efficiency of that workstation could be above 100 percent. That would be great … right? But if the higher-level metric was maximum system Throughput, would running each individual workstation maximize system Throughput, or would it "clog" the system with excess WIP and cause less than optimum Throughput? For this reason, efficiency is not a good metric because it doesn't have a positive effect on the system.

- The metric should be challenging, yet attainable. Suppose efficiency was selected as a metric. Is maximum efficiency challenging and attainable? The answer is no, if you want to optimize system throughput. Because of Step 3 in Goldratt's 5 Focusing Steps (i.e. subordinate everything to the system constraint), non-constraints can never be permitted to run as fast as they can, or they will choke the system with excessive WIP. The excessive WIP extends cycle times, ties up cash unnecessarily and all of the other reasons already mentioned.
- The metrics should lend themselves to trend and statistical analysis and, as such, should not be "yes or no" in terms of compliance. We could definitely trend and perform statistical analyses on efficiency.

The point of this chapter is to demonstrate why it is so important to select the "right" performance metrics. The success of organizations is tied directly to the selection of metrics that drive optimal behaviors. As Goldratt himself said, "Show me how you measure me, and I'll show you how I'll behave." Take a look at the metrics your company is using and see if they pass the metrics litmus test. If they don't, then you have a problem.

REFERENCES

1. Smith, Debra, *The Measurement Nightmare: How the Theory of Constraints Can Resolve Conflicting Strategies, Policies, and Measures*, St. Lucie Press, Boca Raton, FL, 2000.
2. Dettmer, H. William, *Breaking the Constraints to World Class Performance*, ASQ Quality Press, Milwaukee, WI, 1998.

11

Mafia Offers and The Viable Vision

INTRODUCTION TO MAFIA OFFERS

If you're in the consulting world or even just in business making products, did you ever wonder why some companies seem to get more than their fair share of business? We seem to always be looking for ways to get more business and make more profit for our companies. What if there was a method for assuring that your business was able to bring in more work than you ever have in the past? Would that be of interest to you? Success in your business is not a matter of beating the price or service of your competitors, so just what is the key that unlocks your company's success?

This chapter is all about something we call in the Theory of Constraints (TOC) world, a *mafia offer*. No, not going out and hiring gangsters to get you more business, not that kind of mafia. A mafia offer is all about increasing the value of your product or service. It's all about providing an offer that your customers simply can't refuse. Sound interesting?

A mafia offer is an offer that is not difficult to construct, and if it's done correctly, you could reasonably expect an 80 percent offer closing rate. So, let's talk about the basics of mafia offers and how the Theory of Constraints enters into this discussion. Once again, it was Dr. Eli Goldratt who developed the mafia offer back in the 1990s.

Much of what we've written about so far in this book has been about TOC, and we've said many times that there is always an internal or external constraint that is limiting your system's performance when compared to your stated goal. We've told you that the key to profitability is not through how much money you can save, but rather how much you can make. With these facts in mind, one of the keys to the mafia offer is creating a proposal that seeks to maximize your client's profits, by identifying and focusing on your client's system constraint. The mafia offer is intended to remove or lessen the effects of the system constraint and demonstrate more value for your products or services. And for sure, add much more value than your competitors.

The presentation you make to potential clients is really the key to the mafia offer. But having said this, you must know your industry inside and out. You must understand your own capabilities and those of your competitors. Our advice is to follow a very simple presentation format something like the following:

1. Using a very simple PowerPoint presentation, we recommend that you start with an overview of problems within the industry and not the typical company background that most presentations include. They are very boring! Clients want to know right away what you can do for them. Show them the negatives from your industry, and watch to see if heads are bobbing in agreement. Start with things like typical lead times, minimum order quantities, high inventories, on-time deliveries, etc. that demonstrate how the industry's current practices are hurting their business. Stop, every once in a while, and ask them if they are feeling these negatives in their business. What you're doing here is getting them to agree on the problem.
2. The next step is to get the potential customer to agree on the general direction of the solution. Start with something like, "So how do we fix these problems?" We are attempting to describe what life would be like if we had solutions to the problems you just presented. In this step you are developing, with the client, what the basic criteria for the solutions would look like. When this is complete, you should present an overview of your offer first, and then go into a bit more detail for each of the solution criteria. Keep it very simple, but it's important to provide an explanation of the results they will experience. When you've finished this part, ask the potential client if they believe the offer you have presented (i.e. the solutions) will solve the problems they are facing every day. You're looking for agreement here.
3. The next step is to draw a comparison between your company and the competitors. A contrast, if you will. Here is where you offer proof that your solution will work by presenting relevant case studies from clients you've already served. This is the central core of your mafia offer. One key point is that the advantages that you are presenting must clearly be significantly greater than the cost of your product or service.

4. Finally, it's time to close the deal, but keep this simple too. We recommend that you repeat your offer with specifics like, "Our offer will slash your lead times by 'x' days or weeks; improve your cash flow by '' percent; reduce your inventory by 'x' percent while virtually eliminating stock-outs, etc." or whatever you truly believe will happen if they accept your offer. Be bold and ask the potential client what they think. If you get a positive response, you're most probably looking at a contract, so ask them what the next steps are, and then let them tell you what they are.

The key to remember is that a mafia offer order will be something that your competition can't—or is unwilling to—match, at least in the short term.

It is clear to us that there is a significant interest in this subject. And by the way, if you really want to learn more about mafia offers, read *It's Not Luck* by Eli Goldratt and Chapter 22 of the *TOC Handbook* by Dr. Lisa Lang, as both are excellent references. Dr. Lang has a superb consulting business and teaches people how to create mafia offers. We actually borrowed heavily from her work in [1] Chapter 22 of the *Theory of Constraints Handbook* in this chapter.

One question that may be on the minds of many people is, what part of the market segments your company is engaged in can you develop mafia offers for? The simple answer we'd give you is that you could develop a mafia offer for all of your market segments if you wanted to. You might consider things like, which of our market segments deliver the best or worst margins? Or maybe something like, do we have a market segment that has lots of room to grow, versus some that have no room to grow? The fact is, there are lots of things to think about, when you're considering this kind of offer. But the one key point driving the mafia offer is that it's based on having excess capacity.

If you face the following situations on a routine basis, then you are probably ripe to deliver a mafia offer:

1. Your clients order the same product or service from different competitors because they want to be sure one of their suppliers delivers on time.
2. Your customers are complaining about having too much inventory and frequent stock-outs.
3. Your customers complain a lot about late deliveries.

4. Your customers complain about the high cost of products or services.
5. Expediting orders to meet due dates is common with your business and your competitors.
6. Your customers or potential clients are ordering to a forecast.

If you think about what a mafia offer might look like for these situations, you ask yourself, "Can I make an offer that will take away that negative situation for my customer?" If you could come up with a way to counter your potential customer's gripes, do you think your current and potential customers might flock to you with orders? What if you could turn these negative situations into your advantage? Let's look at some simple examples of how you might create a mafia offer.

Let's say you have a client whose business has very high levels of inventory, yet is still experiencing stock-outs of parts on a routine basis. The way we would start off this offer would be with a statement of the known problem, which might look something like this:

We know that in your business you have very high levels of inventory, but we suspect you have very frequent stock-outs of parts that stop your production process. We would like to offer to manage the parts we supply to you. This would mean you don't have to place any orders to us, just send us your usage information every week, and we'll keep you stock levels where they need to be. As a measure of our confidence, we will guarantee no stock-outs. If we do cause a stock-out, we will pay a $500 per day penalty for each day your parts are late. We will also guarantee a 30 percent dollar reduction in the number of parts we hold on site, and you will only pay for what you use, so there will be no more obsolescence losses due to out-of-date parts. We will basically offer our parts to you on consignment.

Another example of a mafia offer might be if there has been a problem with on-time delivery. In this case, we might write the mafia offer a bit differently as follows:

We know that on-time delivery of parts from our company (and our competitors) has been a serious problem for your company. We want to offer you a guaranteed on-time delivery for 100 percent of our parts. Our offer is that if we are late with any order, we will give you a 10 percent discount for each day the order is late. If the order is more than five days late, the entire offer will be free. We are prepared to deliver new parts that we haven't supplied to you previously and will honor the same offer for these new parts.

Let's look at an example of a case where your company manufactures very expensive equipment that customers have been unwilling to purchase because of the price. In this case, you have the option of either leasing it to the customer or charging them so much per hour or day to use it as follows:

> *We know that in today's economy capital expenditures are difficult, so we want to offer you a different kind of deal. Rather than purchasing our J58Z model, we are offering two different possibilities:*
>
> 1. *Your company can lease the equipment for "x" dollars per month.*
> 2. *Your company can agree to pay us "x" dollars per hour only when the equipment is used.*
>
> *We know that this will allow you to keep your cash reserve in the event of serious problems that may arise.*

And finally, what if your organization provides some kind of service, such as a consulting company. How might a mafia offer look here?

> *We know that most consulting companies charge by the day or hour plus travel expenses, but we have a problem with that kind of deal. There is no guarantee that your company's bottom line will improve. While we would still require reimbursement for travel expenses, we would like to offer our services for a percentage of bottom-line improvement. If your bottom line doesn't improve as a result of our services, then you would owe us nothing except travel expenses.*

So, in looking at each one of these mafia offer examples, do you think the company receiving the offer would jump at the chance to sign the deal? We think they would!

In our last section. we gave you several examples of what a mafia offer might look like, but it's important to understand that an expert in TOC can apply the Theory of Constraints to virtually any environment. There are essentially eight different TOC applications that you can use to formulate your mafia offer. For example, if you recognize that your potential client has a problem with shop floor synchronization, or scheduling, then you could apply the TOC technique known as Drum Buffer Rope (DBR). DBR by itself can reduce lead times, inventories, overtime and schedule adherence misses, all with the same level of manpower you have now.

As we said, there are other TOC applications you can use to formulate your mafia offer as follows:

- **Throughput Accounting (TA)**—Traditional Cost Accounting causes managers to make incorrect decisions that typically result in higher levels of inventory (Raw Material, Work-in-Process (WIP) and Finished Goods). By demonstrating this to a potential client, you can convince them that TA will help them reduce these inventories. In fact, you should expect raw material inventories that are 40 percent to 60 percent lower, WIP inventories reduced by 50 percent to 75 percent and finished goods inventories reduced by up to 50 percent!
- **Critical Chain Project Management (CCPM)**—CCPM enables clients to create and execute projects in 25 percent to 40 percent less time than the traditional Critical Path Management (CPM). Clients will be able to meet promised due dates must faster than they ever dreamed possible with typically greater than 90 percent on time completions.
- **Dynamic Replenishment Model**—This TOC application will guarantee 40 percent to 50 percent or less parts inventory with next to zero stock-outs. By simply increasing the rate of parts replenishment based upon usage, you achieve close to 100 percent availability.
- **Strategic Planning**—Using the TOC Thinking Processes, clients can significantly shorten their strategic planning times with much more effective plans developed. One such tool, the Intermediate Objectives Map (also known as the Goal Tree) is a simple to learn and apply TOC tool that always delivers fast, reliable strategic plans.
- **TOC's Five Focusing Steps**—Using the five basic steps of TOC you are virtually guaranteed to significantly improve the throughput of your processes to not only meet current customer requirements, but also grow your market share.

These are but some of the TOC applications you can use to generate mafia offers that will skyrocket your sales.

VIABLE VISION

In this section, I want to talk about a book entitled [2] *Viable Vision—Transforming Total Sales into Net Profits* by Gerald Kendall. To quote one of the reviewers of this book (Patrick J. Bennet, Executive Vice President, Covad Communications), "This book is for anyone responsible for increasing the profitability of their business." When you stop and think about it, aren't we all responsible for increasing the profitability of our businesses? Gerald Kendall is a Principal of TOC International, and a noted management consultant, public speaker and facilitator who has been serving clients worldwide since 1968. He is certified by the TOC International Certification Organization in all six disciplines of the Theory of Constraints, and we highly recommend this book.

So, you may be wondering, just what is this thing called Viable Vision? In its most simple definition, it is the strategy and tactics to achieve, within four years, net profit equal to your current sales. Sound impossible? According to Gerald Kendall and another great consulting professional, Dr. Lisa Lang, it really isn't! Just imagine being able to raise your net profits to your existing sales revenue and to be able to do so in just four years. We won't be trying to explain the details of how to achieve this remarkable level of success in this chapter, so we encourage you to read Mr. Kendall's book or link onto Dr. Lisa's site using this link http://www.viable-vision.com/ to learn more about the details. Although we won't be giving you the details of the Viable Vision, we do want to briefly discuss one part of it in this chapter.

In Chapter 3 of Kendall's book, [2] "Moving from Complexity to Simplicity," he starts with a quote that is worth repeating, "*The more complex the problem, the simpler the solution must be, or it will not work!*" He goes on to explain that "Most organizations deal with complexity by breaking down their organization into functional parts and demanding that each part figure out how to improve itself." Kendall refers to this approach as the *silo approach*. He also explains that cross-functional *battles* emerge as a result. He tells us, "These cross functional conflicts are driven by the silo approach, where the organization measures each silo on improvement independently. If you are a cost center (e.g. procurement, production, engineering), improvement naturally means a focus on cost reduction or greater efficiency within your silo." "In this frame of reference, costs are seen as obeying the 'additive' rule. That is, the costs of each silo, added together, equal the total cost of the organization. Therefore, managers see any cost reduction in their area as 'good' since they see a direct translation to cost savings for the company as a whole."

This thinking comes directly from traditional Cost Accounting's beliefs and teachings, but the fact is, this thinking is flawed. The sum of all localized improvements does not add up to corresponding system improvements. The real key here is *focus and leverage*. We know we must sound like a broken record, but the concept of focus and leverage truly is an important one and really forms the basis of the Viable Vision. The key to driving net profits higher and higher is directly proportional to driving Throughput higher and higher, while maintaining or decreasing Operating Expenses and Inventory. Viable Vision shows you how to do just that!

We encourage you to check out Viable Vision through both of the references I listed earlier, because it will be truly worth your effort.

REFERENCES

1. Lang, Lisa, Chapter 22 of the *Theory of Constraints Handbook*, The McGraw Hill Company, New York, 2010.
2. Kendall, Gerald, *Viable Vision—Transforming Total Sales into Net Profits*, J. Ross Publishing, Boca Raton, FL, 2005.

12

On-the-Line Charting

Most companies who use financial metrics live and die by the quarterly report. Some companies are more inclined to look at it monthly as a decision tool and to get a "feel" for the financial numbers to date. But, even when you track and monitor the financial numbers on a quarterly or monthly basis, there are times when the interval of time seems inadequate. It's disheartening to review a quarterly report and realize you have been losing ground for the last two months and didn't know. Even the monthly interval can be too short to give the necessary "heads up" that a potential problem currently exists. It is much more preferable to have reliable information available on a much shorter time interval. However, compiling and presenting a usable profit and loss statement on a daily basis seems unrealistic.

What if there were a way to compile and analyze financial numbers on a daily basis to help with the decision making—would you be interested? Without a doubt, I'm sure the answer is "Yes!" If you had information available that could tell you where you are *today*, based on business activities as of *yesterday*, then that would be a valuable tool. With this kind of information, you could make the necessary and accurate daily decisions about where you are and where you need to be. It's always much better to know this information as quickly as you can versus waiting until the end of the month, or worse yet, the end of the quarter, before realizing a problem is present.

To view this information daily, enter the on-the-line chart (OTL Chart). The OTL Chart is a simple concept that is based on the rules of Throughput Accounting (TA) to allow the user access to the listed information available in real time. Setting up an OTL Chart is very easy. First, you need to have an estimate for the Operating Expense (OE) on a monthly basis. In other words, how much does it cost you to keep your business running every month? The beauty of the OTL Chart is that it does not need to be accurate down to the dollar or the penny. It is understood that there can be fluctuation in the monthly dollar amounts. The OTL Chart is strictly a compass to keep you heading in the right direction and notify you of potential issues when they happen.

There are a couple of ways to get these numbers. You could take yearly expenses and divide by 12 to get an estimate of monthly cost. Or you could take monthly OE and divide by the number of days in the month to get a daily cost. What we want to end up with is an estimate for the daily OE being accumulated. Once you have that number, you can plot it on a graph using Excel. Let's look at a simple example.

Suppose your business that had a monthly OE of $30,000, and you want to plot this number for the month of June. If we take the $30,000 and divided by 30 days, then the estimate is about $1,000 per day of OE. We would plot this in Excel as a cumulative number. That is, Day 1 equals $1000, Day 2 equals $2000, Day 3 equals $3000, and so forth. Figure 12.1 gives an example of what this chart might look like.

With the information plotted, this graph shows the daily cumulative totals for the entire month. It is possible that the OE numbers could

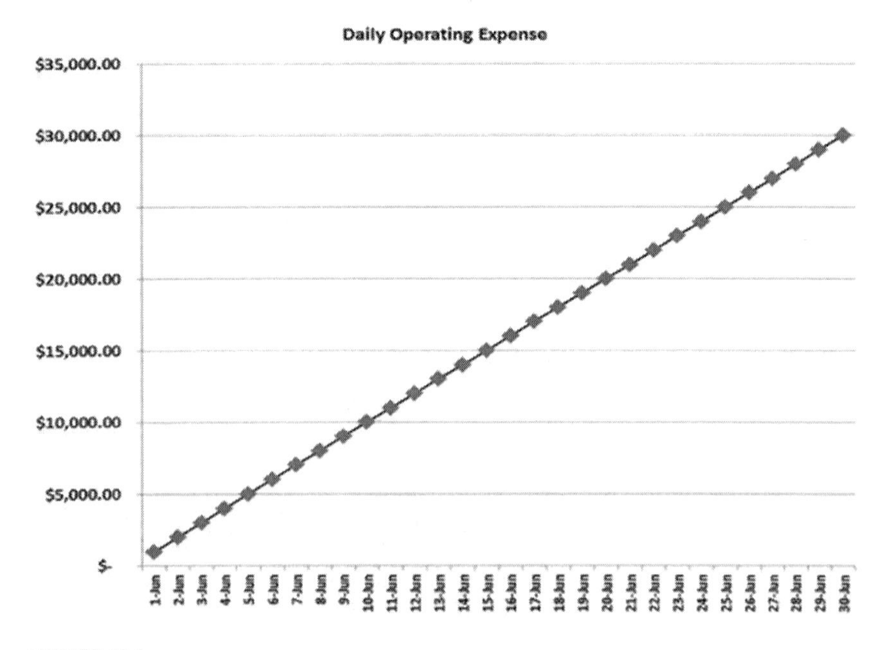

FIGURE 12.1
Example of What an On-the-Line Chart Might Look Like.

change in any month because of things like some employees quitting or new employees being added. If the variance is higher or lower, then adjust the numbers accordingly. However, what we are really trying to establish here is the view of the OE from the ten-thousand-foot level and not necessarily the day-to-day changes. Think big picture and not finite detail.

With the OE line established, we now want to the collect the Throughput data. Remember: This only works if we are collecting and reporting Throughput in accordance with Throughput Accounting rules. As such, Throughput is calculated as product selling price minus totally variable costs (T = SP – TVC), while Net Profit equals Throughput minus Operating Expense (NP = T – OE). As you probably already know but, let's refresh anyway, TVC is any cost associated with the product. This will include raw material, sales commission and shipping charges. Remember, labor charges *are not added into this number.* Labor costs are part of the operating expense calculation.

Suppose the product we make has a selling price of $90.00 and a TVC of $25.00. That means for each product sold, we have a throughput value of $65.00 ($90.00 (SP)–$25.00 (TVC) = $65.00 (T)). With this information, we now know that to break-even on the OE, we must make 15 or more products per day. With 16 or more products per day, we start to make a profit. It is interesting to note that the Cost Accounting rules will tell you that you are making a profit with each product sold. TA counters with the realization that profit does not and cannot begin until the 16th product is made and sold. It is a vastly different financial concept to think of product pricing and product margins using TA.

If you track the daily Throughput from the system and calculate Throughput correctly, you should have a pretty good idea where your company stands right now. Figure 12.2 shows the impact of Throughput to OE after 17 days of tracking.

Using Throughput Accounting and looking at this chart on a daily basis can give a General Manager or Department Manager accurate and useful information. The analysis is simple. If Throughput is tracking *below* the OE line, then you aren't making enough money to cover the OE expense. The management team can determine the issues causing the lower than necessary Throughput and initiate corrective actions to bring the Throughput line up. If the Throughput line is tracking *above* the OE line, then you are making a profit, and possibly no actions are required.

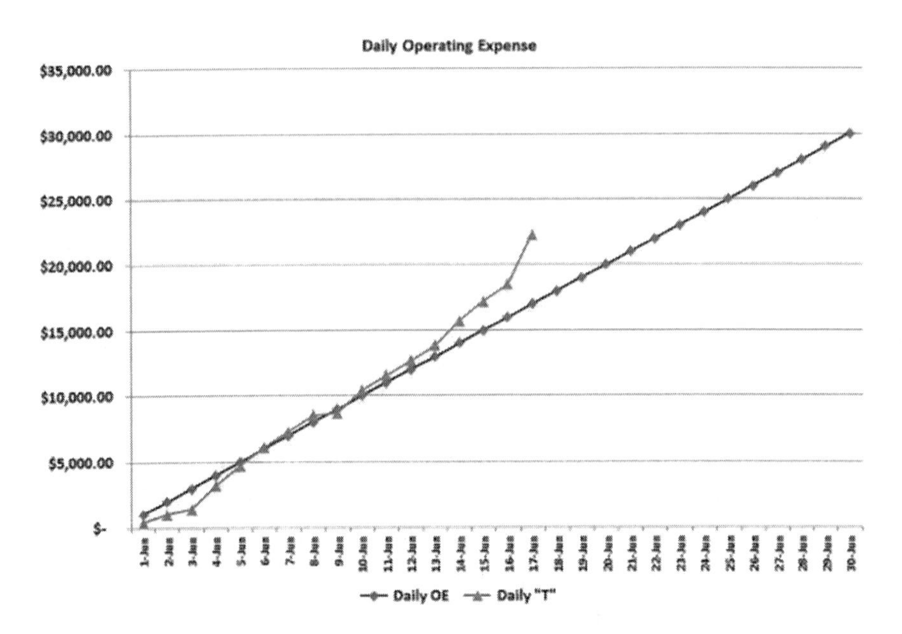

FIGURE 12.2
The Impact of Throughput to OE after 17 Days of Tracking.

The OTL Chart is a great tool to track and monitor the OE and T on a daily basis. It gives you a good "Kentucky wind age" analysis to determine where your company is right now. There will be no need to wait for the month end or quarterly reports to figure out what happened, either good or bad. Using the OTL Chart, you can have a very good estimate of how the company is performing and make any necessary corrections to get back on track for revenue and on-time delivery.

13

Active Listening

INTRODUCTION TO ACTIVE LISTENING

In this chapter, I want to share a very powerful technique I refer to as *Active Listening*. This chapter is directed at companies where the workforce morale is not good, and for all of those companies who need a rapid improvement in Throughput. I've been using this "technique" for the last decade, and every time I do, rapid improvement in Throughput has been the end result.

Many companies today praise themselves for how well they "involve" their workforce in their improvement efforts. In fact, if you go into many companies, you'll probably see a wall of pictures that support the contention that "our people are involved." And although I love seeing this "gallery of involvement photos," often they are just photos and not what is actually happening. So the question becomes, just how involved are the true subject matter experts (SMEs) within your company? This depends on what your definition of "being involved" really is and who you believe are your true SMEs.

Involvement in many companies is simply participation on improvement teams. But in my mind, simply participating on a team is not enough to generate significant improvement. Why not? The best way to answer this question is through a simple case study, based upon a consulting engagement that I was fortunate enough to lead.

This company was in the Aviation MRO (i.e. Maintenance, Repair and Overhaul) industry and was a contractor to the Department of Defense (DoD). This company, by contract, was required to supply a predetermined number of aircraft every day, and if they didn't, they were assessed a significant financial penalty, based on the number of aircraft that they failed to deliver. This company was struggling to meet demand, and the results were getting worse. So much so, the corporate office replaced the site leader, in hopes of turning this company around.

In addition to the financial losses for missed aircraft availability, this company was paying a huge amount of money for mandatory maintenance overtime in an attempt to "right the ship," so to speak. One of the consequences of this mandatory overtime, which had been in place for months prior to my arrival, was extremely low workforce morale. The more overtime the maintenance workforce was mandated to work, the lower the morale became. Call-ins and absenteeism were high as well, which were directly the result of this constant overtime.

Upon arriving to this site, I met with the new site leader to discuss his issues, and it was clear to us that he was frustrated. And while the outgoing leader's management style was command and control (i.e. do it my way!!), the new leader believed in listening to fresh new ideas. When I asked him if he was ready to involve his people, he replied that they already were involved. When I asked who his SMEs were, he gave me a list of technical people (i.e. Engineers, Supervisors, etc.) on site. I responded with, "So, these are the people that physically maintain the aircraft?" His response was, "Well no, but they are the experts." I just smiled back and said, "No they aren't!" I explained to him that the true subject matter experts are the people that maintain the aircraft. The Mechanics, the Avionics Technicians, the Quality Assurance (QA) folks, the Maintenance Control people, the flight-line workers and Logistics workers. The look on his face was priceless, like he had just had an epiphany of sorts. I then explained my version of employee involvement, *Active Listening*.

I explained to him that if he wanted to rapidly turn around his results, the first thing he needed to do was form a team comprised of only SMEs. This team needed to be made up of all of the maintenance-related disciplines, but that membership needed to be completely voluntary, and wherever possible, it needed to be the informal leaders of the workforce. This was a union environment, and the new site leader was somewhat concerned that the union wouldn't agree with this method.

I then explained the central concept of Active Listening, which is that the managers would not just listen to the core team's ideas. As long as their ideas and solutions didn't violate any safety policies, customer or company policies or contractual obligations, then their improvement ideas must be implemented exactly as stated by this core team. I further explained that this would be difficult, if not impossible, for some of his managers and supervisors to do, but that it was absolutely necessary for success. The site leader's response seemed to be somewhat positive, but also somewhat disbelieving that this approach would work. He did, however, commit to us that he would support this approach.

Several days later, our first core team meeting took place. In this first meeting, the maintenance process was mapped out. I wanted to make certain that everyone understood exactly how their maintenance process was currently working. This proved to be a valuable learning experience for some of the team members, because they got to see first-hand how their work impacted the flow of aircraft through the maintenance process and through the maintenance hangar.

Using the teaching method outlined in previous chapters, I then presented the basics of the Theory of Constraints (TOC) to this core team. Without exception, everyone understood this new concept. As a team, we then identified the system constraint to be all of the actions required to be completed (e.g. approvals, parts availability, etc.) before maintenance work could begin on the aircraft (i.e. full-kitting). I then gave them training on the Interference Diagram (ID) and asked the group for solutions to each of the interferences they had identified.

There was skepticism that management would implement their solutions, but the site leader assured them that their ideas and solutions would be implemented exactly as presented, as long as no safety policies, company policies or contractual obligations were violated. As this first meeting came to a close, we asked the members to go solicit additional ideas from their co-workers.

New ideas came from everywhere, and most of them were implemented exactly as stated. The core team itself was responsible for deciding which shop floor ideas would or would not be implemented. You could see the workforce morale changing, improving and growing. I met with this team twice a week and began a brief newsletter of sorts, to communicate the actions of the core team. The results came swiftly, and within two to three weeks, the aircraft availability targets were being consistently met. In three to four additional weeks, all mandatory overtime was stopped. The workforce morale rocketed upward, and availability targets were met at an even better rate. All of this improvement came as a result of Active Listening, the identification of the system constraint, and focusing our improvement efforts (the core team's solutions) directly on it.

The improvement continued for the next few months, and when it came time for renewal of the maintenance contract, we used some of the original core team members to present what they had done to the DoD contract specialists. Needless to say, the contract was renewed without objection. This facility was visited by other maintenance sites within the same company to see why it was functioning so very well.

Active listening has always worked for me, no matter what the industry type. I have used it to improve manufacturing facilities, healthcare facilities, MRO companies, etc. simply because of its implementation simplicity. The basic steps for preparing to implement this improvement method are as follows:

1. Visit the site needing improvement and perform a "system walk-through" just to acquaint yourself with the various functions and processes.
2. Take the time to stop and talk with the people who actually do the work, the true subject matter experts.
3. Make a list of problems that you see in the facility. Things like inventory level, where the bottlenecks (constraints) are located, etc.
4. Determine what type of scheduling system is being used. Determine if the system is based upon "push" or "pull."
5. Meet with the facility leadership and ask them what they see are the problems that exists within the facility. My guess is that leadership's list and the list created with help of the SMEs will be totally different.
6. Determine how truly "involved" the SMEs really are in the facility's improvement effort. That is, find out if they are simply team members or are they actively involved in the improvement effort. My guess is that they are probably token members of improvement teams.
7. Meet with the senior leadership of the facility and explain how active listening actually works. That is, form a core team of your true subject matter experts, explain that you are looking for improvement ideas and that, as long as the ideas don't violate any safety regulations, customer-specific requirements or company policies, the idea WILL BE IMPLEMENTED exactly as presented. The most effective way to communicate this message to the core team is to have the most senior leader of the facility present Active Listening directly to the facility's core team.

8. Develop a method to post the status of improvement projects for everyone in the facility to see and review on a regular basis.
9. Celebrate your successes!

As I said earlier in this chapter, involvement in many companies is simply participation on improvement teams. But in my mind, simply participating on a team is not enough to generate significant improvement. Active Involvement works!

14

Is Change Really Necessary?

INTRODUCTION

The honest answer to this important question as to whether change is necessary is that "it depends." It depends if the change you are making is really necessary, or are you just changing things because you can? It depends on whether or not the change is associated with a systems constraint, or is the change a non-constraint?

Let's talk about "unnecessary change" first. Sometimes, change just for the sake of change can have destructive outcomes, no matter how good the intentions are. Unnecessary change is most commonly associated with organizations that are working in isolation from each other, with no real "focus" on the overall goal of the company. Each individual organization has determined some predefined goal(s) that they want to accomplish, and they set out to do so. Sometimes they do this without any real understanding of the overall systems effects that the proposed change might have on another organization.

As an example, suppose a sales organization wants to increase sales without a concrete understanding of the internal capacity of the manufacturing organization. More sales, without the necessary capacity, will be very destructive to the manufacturing organization. There will be increased late orders, longer lead times and unhappy customers. So, an improvement in one organization can have a very destructive effect on another organization. What started out with good intentions quickly became a big problem for the entire company.

Now let's talk about necessary change. Any change that can move the company closer to its overall goal (make money) is probably a very good change to make. Any recommended changes brought forward can be evaluated with a quick and effective litmus test. Ask yourself, "If I make this change, will Throughput (T) go up?" You can also ask "Will Operating

Expense (OE) stay the same, or go down?" Or, "Will Inventory/Investment stay the same or go down?" If the answer to any of these questions is no, then it is probably not a good or necessary change to make and should be shelved until another time.

However, making necessary changes does require some accurate information. First, you must know where the system's constraint currently resides. Second, is accurate (probable) information known about where the constraint will move next? If the system constraint limits the system output, then any improvement of the constraint will improve Throughput through the system. The first litmus test has been passed! If you spend your time and resources "focused" on anything except the constraint, you will miss the opportunity for maximum "leverage."

If you have a good idea where the constraint will move next, then the necessary planning can be undertaken to deal with the next constraint. This sequence of finding and fixing is exactly the same as the "piping diagram" that has been referred to many times in this book. Find and fix the first constraint, and move to the next one. This sequence allows you to make the necessary improvement because you "must," and not just because you "can."

GETTING BUY-IN FOR A CHANGE

Sometimes getting the necessary buy-in for the changes you want to make can be a difficult process, but not impossible. In the TOC Applied Systems Thinking course (Jonah Course), there is a segment (module) dedicated to this process, which provides some useful guidelines to implement change.

In general, some people tend to resist someone else's ideas for change. It falls under the "not invented here syndrome" and can, at times, be troublesome to overcome. However, there is also some simple and powerful psychology involved, and it helps if you understand that

- Some people have a very powerful intuition in areas where they have experience.
- Some people don't recognize the need for change.
- Some people don't always understand what needs to be changed or why.

- Most people want to feel comfortable that the change is likely to succeed.
- Most people want to understand how any change will impact them.

If you take the approach of just presenting "your idea," it will be a challenging effort to succeed. However, if you ask for and accept input(s) from others, your ideas will have a much greater probability of success. Allow other people to modify or even criticize the solution, then ask them to help make the corrections; in other words, ask them "What would you do differently?" Always assume the other person has a good point, even if they have not presented it well. Listen first to understand what the person means, and not just what they are saying. And by all means, never make the other person look bad. You need to always show how the solution leads to *their* benefits and addresses *their* problems. If you give them the opportunity to help design the solution, the chances of their buy-in will be almost 100 percent.

While some people will resist change, in most cases there is at least one person who does not resist, *the person who invented the idea.* First and foremost, you want to seek to create ownership of the idea(s) that you want to see implemented. It is not uncommon that the emotion of the idea's inventor will provide a very powerful platform to guide other people's energy toward supporting an idea. By allowing other people to modify your ideas, you create a situation where the solution(s) can become "other people's idea(s)," and not just yours. In essence you have enacted the "Socratic Method" that allows others to participate. What is very important at this stage is that you must not rush to reveal your answer. You should always allow the person time to digest your ideas and reach the same conclusions on their own. When the "new" idea becomes "their" idea, you have successfully used the Socratic Method to create the necessary ownership. When other people own the idea (solution), they will most likely make it happen in very short order.

Logic can be one of the most powerful tools we use to gain a consensus for ideas. Logic, both necessity and sufficiency, can be used to show how something systematically will help to solve a problem, reach an objective or overcome an obstacle. As powerful as logic is, emotions are even more powerful. When provoked and pushed to the limits, emotional resistance will block even the most solid logic. Emotional Resistance to a good idea can come in many forms, but the three most prevalent are

1. Showing the person responsible that they are wrong (making him or her look bad).
2. Acting as if *your* solution is the answer to the world's problems (it's probably not, so don't pretend that it is).
3. Let it be the others person's idea … it's OK!

The scenario that you really want to end up with is a situation where you can help others recognize the existence of the problem and/or the need to change. The starting position cannot be one of "you" against "them," but rather strive for a position of "you" *and* "them" against the problem—not against each other. If you approach it in this manner, you will enable others to see a way out of the problem or a solution for the problem. A solution developed with others is a solution that leads to everyone's benefits—in essence, the win-win scenario. Remember, there is no useful solution except for the win-win. Anything else is just a win-lose.

In your desire to implement change, you will likely encounter some other categories of people. Through time we have narrowed these down to three categories. These categories are not based on job functions or organizational titles. These categories of people can exist anywhere, up and down the organizational chain. It is highly probable that once you understand these categories, you will know instantly when you run into one of them.

The three categories of people are

1. **Directly Responsible Person (DRP)**

   The type of person is affectionately known as the DRP. This is the person who is tasked with responsibility for the core problem, or the area that you are considering for change. They very clearly understand the subject matter, but they are also extremely sensitive to (and very often tired of) being blamed for all the problems. What this type of person wants more than anything is a way out of the problem. These types of people usually suffer from a martyr complex and will feel directly attacked, even if they aren't directly responsible.
2. **Intimately Involved Person (IIP)**

   The IIP clearly understands the environment where change is needed. It is possible this person is the next level up in the organizational chain, but it is also possible that they exist in areas outside of the organization. If correctly situated within the organization, the IIP can be very important for gaining consensus for your new ideas and change. They are a great person to have on your team.

3. **Outside Person (OP)**

   The outside person is usually totally unconcerned or unaware of Undesirable Effects (UDEs) that exist. They usually sit at the higher level of the organizational structure and perhaps even at a corporate level. They are, for the most part, disconnected with the realities of the lower organizational structure. The connection between necessary actions and the implied benefits usually isn't obvious to them. However, it is highly probable that you will need their cooperation for the intended solution.

By understanding and looking for these three categories of people, you can learn to temper your buy-in approach, either up or down, to get the buy-in and consensus you need. Good luck with your approach.

15

TOC in MRO

INTRODUCTION

In the past, I have worked for and consulted for Maintenance, Repair and Overhaul (MRO) contractors within three of the Department of Defense (DoD) branches, including the Army, the Navy and the Air Force. My first engagement in this area was for a contractor who worked for the US Army. This contractor was hired to maintain a variety of helicopters, and my role was to help the contractor improve the throughput of helicopters through their preventive (and reactive) maintenance processes. I have also consulted for contractors to both the Navy and Air Force, helping them improve their MRO on things like jets, jet engines and other helicopters. In this chapter, I want to present several examples of how I helped these groups attain significant improvements.

In the case of the contractor who maintained engines, when an engine was due for service, the contractor had to replace it with a rental engine while the original engine was being maintained, repaired or overhauled. As part of the contract, the contractor had to pay for the rental expense if the MRO completion time exceeded the contract limit, which it typically did by a wide margin. I was called in to help this contractor reduce the MRO time on the engines.

I used my customary approach, in that I selected a core team of the true subject matter experts (i.e. the front-line workers) and had them map the process to identify the location of their system constraint. In doing so, the team identified two constraints within this process, and both of them involved lengthy approval processes with lots of wait time imbedded in them. And while the paperwork was being approved, the cost of the rental engine kept accumulating. It wasn't the engine repair time that caused the extended cycle time problem at all, but rather the time waiting for approval signatures. This was a classic example of policy constraints causing the MRO time to exceed the contract limit, resulting in excessive rental costs. Once these constraints were identified, it was clear what had to be done. The real question that had to be answered was, "Why were the approval process cycle times taking so long?"

We looked at a lot of data, including the email trails, the vehicle to transmit the approval paperwork. What we found was that because the manager responsible for approval had so many other functions to perform, he would send out his paperwork usually only one day of the week. And when he did so, he sent it out in "batches." The effect of him "batching" the approval paperwork was exactly the same effect as a python trying to swallow a pig! The python can do it, but the process is slow, and the pig moves through the python's body at a snail's pace. Yes, it eventually is digested, but it takes much longer than if the python had eaten the pig one bite at a time. Batching encumbers a process by extending the overall cycle time of a process, and the approval paperwork process was no exception. So, what did we do to "fix" this problem?

We watched the process of entering the data into a database by the Engine Manager, and it was clear that it was a lengthy process for him. It was also clear that by hiring a data entry person, the Engine Manager could be freed up to perform other important functions. The Costs Accountants told us that they were not permitted to hire anyone, but that we could run a three-week study, which we did. During that three-week trial, the approval paperwork jettisoned through the process and the rental engine time and expense decreased significantly. Problem solved! Right? We wish it would have been that easy.

The Accountants would not approve a permanent slot for data entry because it was too much of an expense! Think about this decision for a minute. If the cost of the rental engine was $75/hour, and we were able to reduce the rental engine time from 60 to 20 hours, well you can do the math. In the Cost Accounting (CA) world, the focus is on cost savings because they believe that the key to profitability is through saving money. In the Theory of Constraints (TOC) world, the key to profitability is through making money, and the key to making money is by increasing the throughput. And the key to increasing throughput is by focusing the improvement effort on the system constraint. But then again, not everyone sees it this way. This is an example of how a very profitable solution to an existing problem was overruled by Cost Accounting's belief that the key to profitability was through saving money.

As I've written about in an earlier chapter, the Theory of Constraints version of accounting, referred to as Throughput Accounting (TA), demonstrates quite nicely that the key to profitability is through making money. The difference between these two approaches (i.e. CA vs. TA) toward profitability is radically different. I'm going to demonstrate this difference through another mini case study from the same parent company as before, but a completely different environment. You will recall that the initial case study was from an organization responsible for making sure that jet engines were maintained, repaired and overhauled. This case study involves the maintenance, repair and overhaul of helicopters that we discussed in our chapter on Active Listening.

When I arrived at this site, I found quite a chaotic environment. The client had to supply "x" number of helicopters every day or they would receive a substantial dollar penalty for each aircraft that fell below "x." Because they were significantly below "x," the financial penalties were substantial. Additionally, they were on mandatory overtime to the tune of 12 hours per day, which was further eroding their profit margins. Because of the sustained mandatory overtime, the workforce was exhausted, and the morale was absolutely terrible. Furthermore, because they were not synchronizing the flow of aircraft into and out of the maintenance hangars, there were significant numbers of aircraft tied up in work-in-process (WIP). The airfield site leader and maintenance manager had just been changed out by their corporate headquarters, and the new team was desperately looking for help.

I met with the site leader and his team to better understand how serious their problems were, and as the meeting progressed, it was clear that action had to be taken sooner rather than later. I also met with the shop floor people, and it became clear that they were not engaged at all. They were tired of working overtime every day and felt that they were not appreciated or respected. They were happy to see a change in leadership, because the previous leaders were very much into command and control management. It was so bad that the workers were out looking for lower paying jobs just to get away from this site.

My first action was to explain to the leadership team that if they wanted to turn this site around and make it profitable, then they had to significantly improve the throughput of helicopters through their scheduled and unscheduled maintenance processes. I explained that to do that successfully, they had to engage their total workforce. I then told them that the only way to engage their workforce was through something I refer to as *Active Listening*. As explained in an earlier chapter, Active Listening is achieved by forming a core team of a cross section of the hourly workforce, soliciting ideas and solutions from them and then implementing the solutions exactly as presented, provided that there was no violation of safety, regulatory, company or contractual obligations. Because the leadership team had no other recourse, they agreed to do this.

The team was formed, and our first order of business was to map the process. When this was completed we gave the core team some training on the Theory of Constraints and Goldratt's 5 Focusing Steps. Our next order of business was to create a strategic plan for improvement, and the tool we used was a Goal Tree (Figure 15.1). In this exercise, we used a team comprised of both management and members of the hourly core team.

The development of this Goal Tree proved to be a very worthwhile exercise for several reasons. First and foremost, for the first time ever, leadership and hourly employees were working side by side to improve the overall maintenance system within this facility. Secondly, the completed Goal Tree was used to assess the status of the maintenance system in general, which allowed the team to prioritize and focus on the most important areas for improvement. Thirdly, the combined leadership and hourly team was able to create a strategic improvement plan. The team elected, as a starting point in their improvement initiative, to use their newly constructed Goal Tree to assess how their organization was currently performing in their unscheduled maintenance area. The reason we started with unscheduled maintenance was because this contractor was not supplying enough aircraft in time to support the training needs of their customer. And as they discovered, their performance was not acceptable. The team used a color-coding system to assess this area, which worked as follows:

1. If the activity was in place and functioning well, the team was instructed to color it green.
2. If the activity was in place, but not functioning well enough, the team should color it yellow.

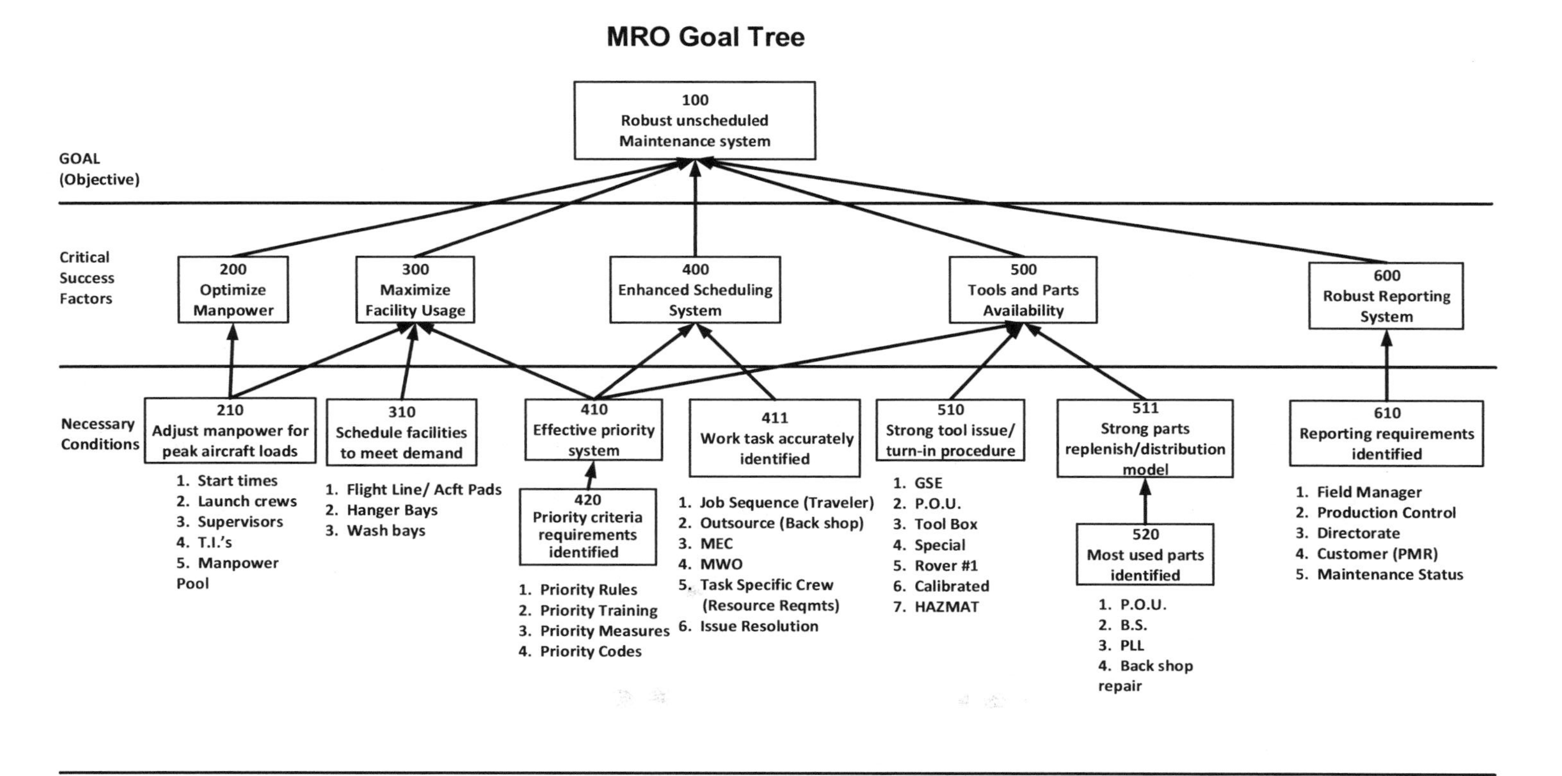

FIGURE 15.1
MRO Goal Tree.

3. If the activity was either not in place or not functioning for its intended function, then the team should color it red.

As can be seen in their completed assessment (Figure 15.2), this team had a significant amount of work to do to turn around their unscheduled maintenance efforts. I instructed the team to start with the red entities, correct them first, and then develop plans to turn them into yellow or green entities. They would then work on the yellow entities to turn them into green ones. The team was surprised to see that there were no green entities. We then turned our attention to the scheduled maintenance area. As they had done for their unscheduled maintenance system, the assembled team developed a Goal Tree to improve their turn-around time for scheduled maintenance aircraft. Figure 15.3 demonstrates the activities required to achieve their goal of "Timely Turn-Around of Scheduled Maintenance Aircraft to Satisfy Customer Needs." The team then used their Goal Tree to assess how they were performing using the same three-color coding system, as is demonstrated in Figure 15.4.

The next step for the team was to develop improvement plans for both their scheduled and unscheduled maintenance systems. In the final analysis, the key activities for these two critical areas were developed per Table 15.1. Table 15.1 is their high-level plan, which the team used to develop a more detailed plan.

One of the most popular tools used by the team was the Interference Diagram (ID) presented in an earlier chapter. The purpose of the ID is to identify those "things" that are "blocking" attainment of your goal. The team decided that what they really needed (i.e. their goal) was "More Wrench Time," and by identifying those things that are stealing away their wrench time, the team believed that significant progress could be made on their aircraft turn-around time in unscheduled maintenance. The team developed their interference list and then "guesstimated" the amount of time they were losing as a result of these interferences. Figure 15.5 is the ID developed by this team. There were several areas that stood out in the ID, and these became focal points for their improvement efforts. One area in particular, "Finding and signing tools in and out," accounted for 1.5 hours per day of lost wrench time. Another area was their safety and fall protection equipment. It seems that they had two hours of lost wrench time either searching for or waiting for the equipment. One by one the team developed solutions to these interferences whereby they either eliminated them altogether or found ways to reduce the lost time.

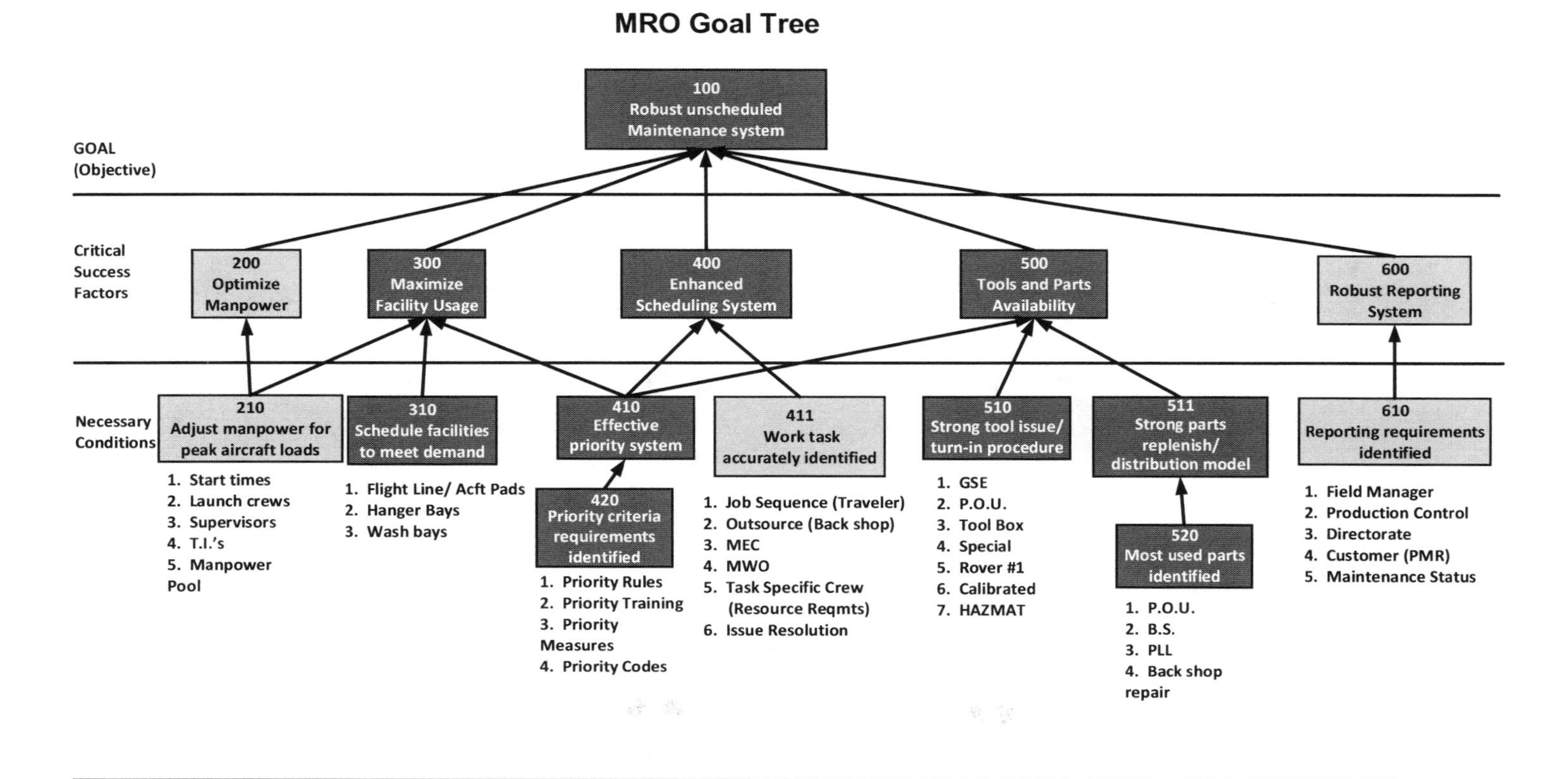

FIGURE 15.2
MRO Goal Tree after Assessment.

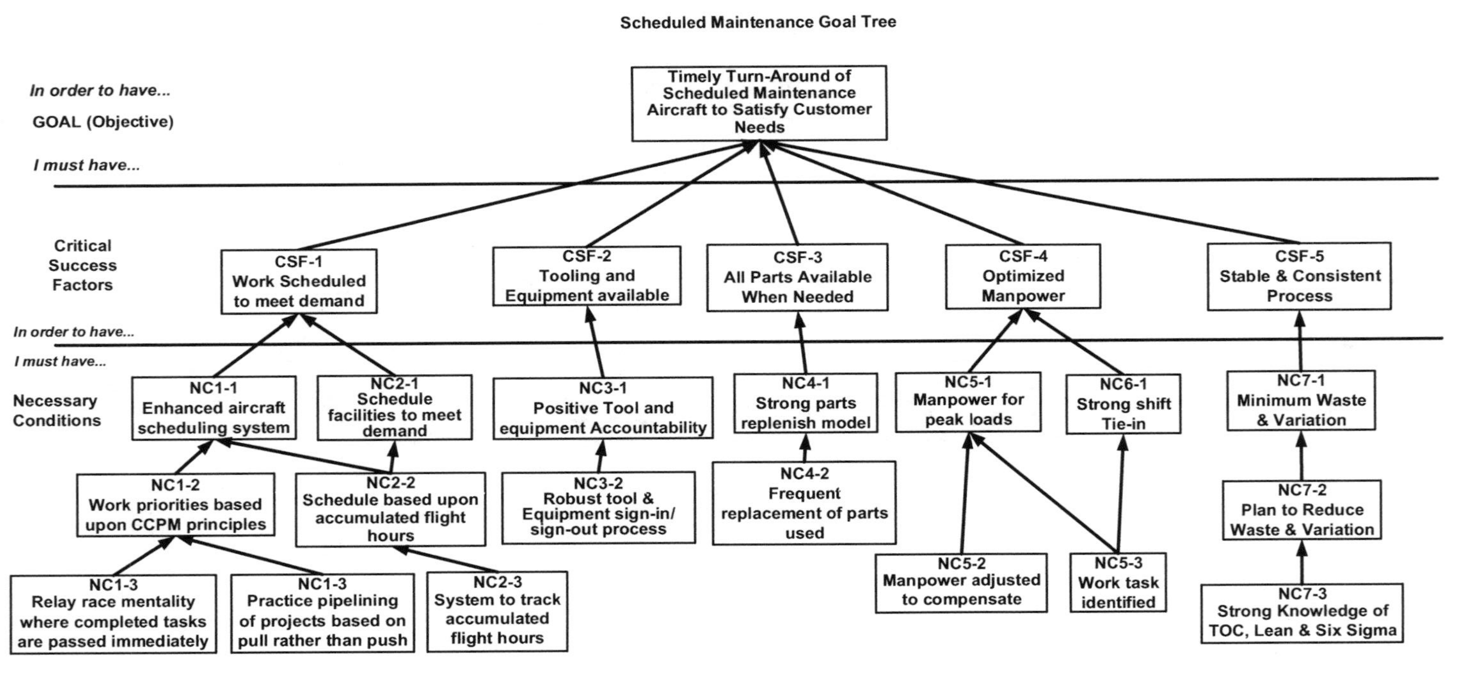

FIGURE 15.3
Scheduled Maintenance Goal Tree

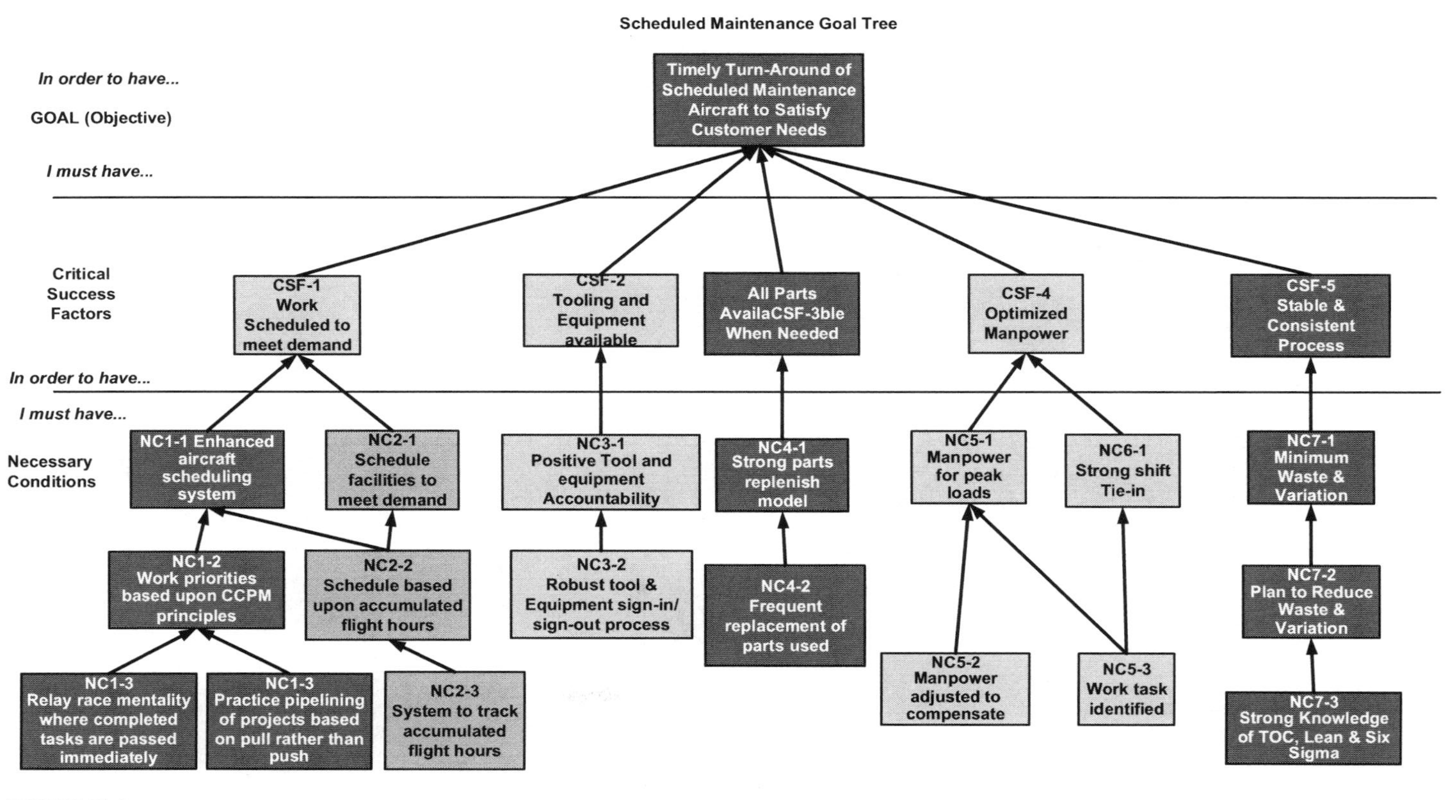

FIGURE 15.4
Scheduled Maintenance Goal Tree after Assessment.

TABLE 15.1

Scheduled Maintenance Goal Tree

Area of Concern	Scheduled Maintenance	Unscheduled Maintenance
Parts/Supply Shortages	TOC Parts Replenishment System	TOC Parts Replenishment System
On-Time Completions	Critical Chain Project Mgt. (CCPM)	Multiple Drum Buffer Rope (MDBR)
Tools/Equipment Available	Project Kitting	Project Kitting
Waste/Variation Reduction	TOC-Lean Six Sigma	TOC-Lean Six Sigma
Manpower Availability	Manpower "Floaters"	Manpower "Floaters"

Figure 15.6 is a pie chart that summarizes the key interferences the team found. What stands out is that because of all of these interferences, the available work time was only 45 percent of their day.

The improvement to unscheduled maintenance throughput was immediate and swift, and within two weeks three significant changes had occurred:

1. The number of daily available aircraft increased above "x" so that penalties were almost completely avoided. So much so that they actually had spare aircraft available and dollar penalties dropped to nearly zero!
2. All mandatory overtime was stopped, with the reduction in overtime dollars dropping to levels not seen in previous years.
3. The morale of the workforce rocketed upward to levels not seen in years, if ever.

DRUM BUFFER ROPE

One of the other problem areas in unscheduled maintenance was their lack of synchronization due to their inability to predict how many or when the next aircraft would need maintenance. The Theory of Constraints offers a solution to this dilemma. A scheduling system known as Drum Buffer Rope (DBR) is the "tool of choice." So just what is DBR? DBR is TOC's method for scheduling and managing operations, which consists of three different parts:

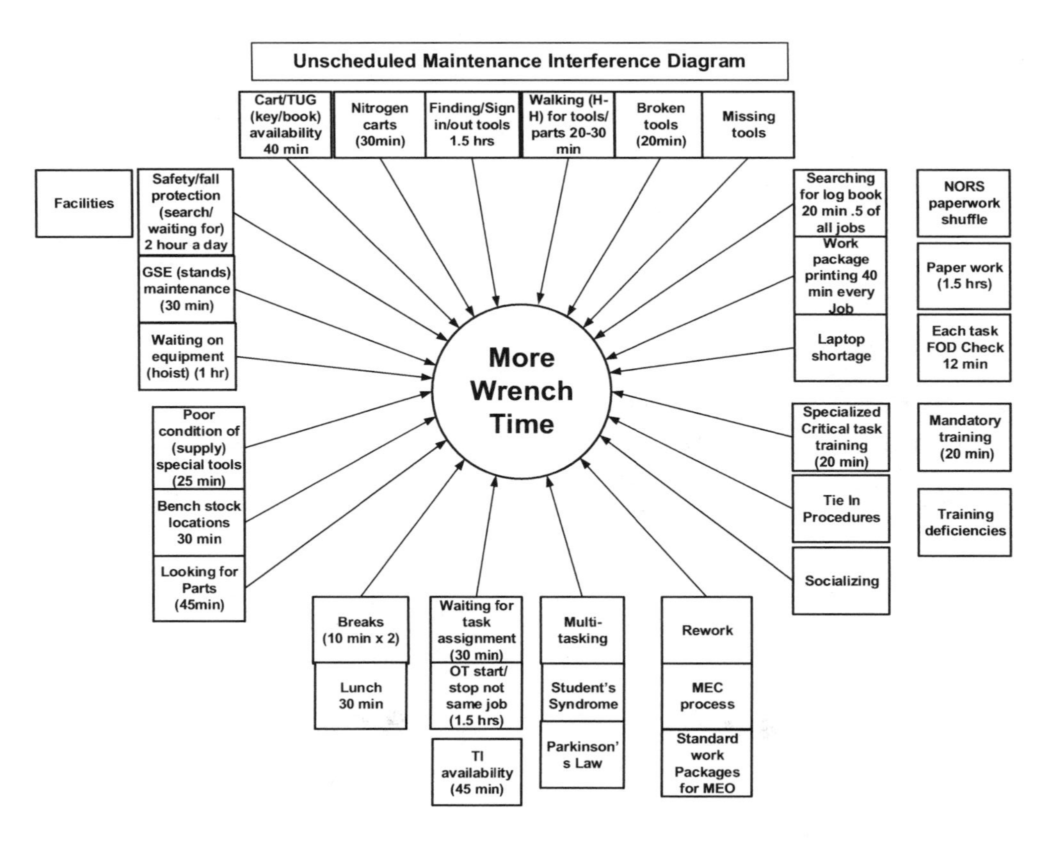

FIGURE 15.5
Interference Diagram Developed by Team.

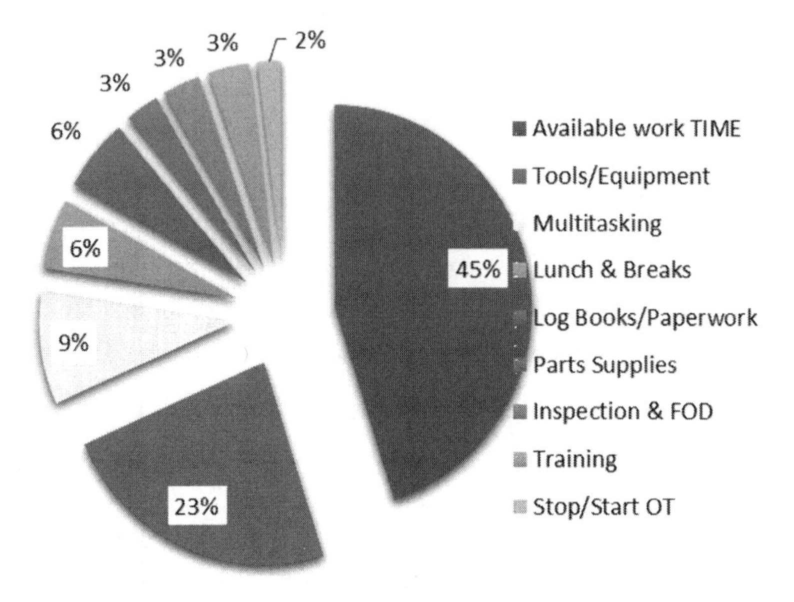

FIGURE 15.6
Pie Chart That Summarizes the Key Interferences Found by the Team.

1. The Drum—generally the constraint or the Critically Constrained Resource (CCR). Sets the pace for the operation and is usually the slowest operation.
2. The Buffer—usually expressed as time, protects the system from variation and uncertainty.
3. The Rope—A signaling mechanism or communication device to control the release of materials coming into the system to match the constraint's consumption.

DBR is designed to satisfy customer demand by delivering all orders on time. It will move all WIP through production/maintenance as fast as possible and will also reveal hidden capacity. What DBR does not do is concern itself with local efficiencies, and it does not encourage "Build to Stock" while more important orders wait.

Figure 15.7 is a graphical illustration of how this team saw DBR working at their maintenance facility. In a nutshell, the aircraft needing maintenance lands on the Flight Line where it is inspected and repaired, if possible. For those aircraft that can't be repaired on the Flight Line, they are sent to the maintenance hangar for inclusion into the DBR scheduling system. The "Drum" sets the pace for material release, and as

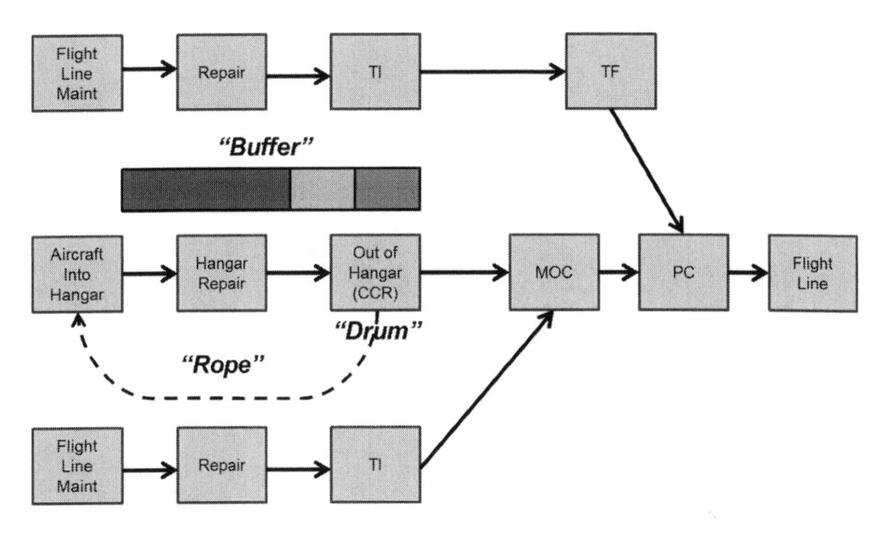

FIGURE 15.7
Graphical Illustration of How This Team Saw DBR Working at Their Maintenance Facility.

material is released from the "Drum," a signal is sent back (the "Rope") to material release to introduce more into the system. In this case, the material to be released is the next aircraft in line to be worked on. The "Buffer" is a stack of work in front of the "Drum" representing time, and the "Drum" should never be starved of work. While Figure 15.7 is the classic layout for DBR, we developed another version of DBR as is depicted in Figure 15.8. We refer to this design as "Multiple Drum Buffer Rope (MDBR)" simply because we saw each Maintenance Work Bay as a separate Drum.

This layout illustrates the triage that takes place on the Flight Line, that is, an examination of the aircraft to determine the problem to be fixed. If it can be fixed in triage, then it is repaired and released. If it must go to the maintenance work bay, a call is made to Production Control whereby it is placed into the work queue. The work queue is designed in such a way that the repair would be made in the "next available work bay," which is the purpose of DBR. That is, to schedule and manage the aircraft based on work bay availability. This layout made it much simpler for Production Control and the Maintenance Supervisor to work together to maximize output of the total unscheduled maintenance hangar. The turn-around time for aircraft improved dramatically, and the DoD customer was very happy with the results achieved.

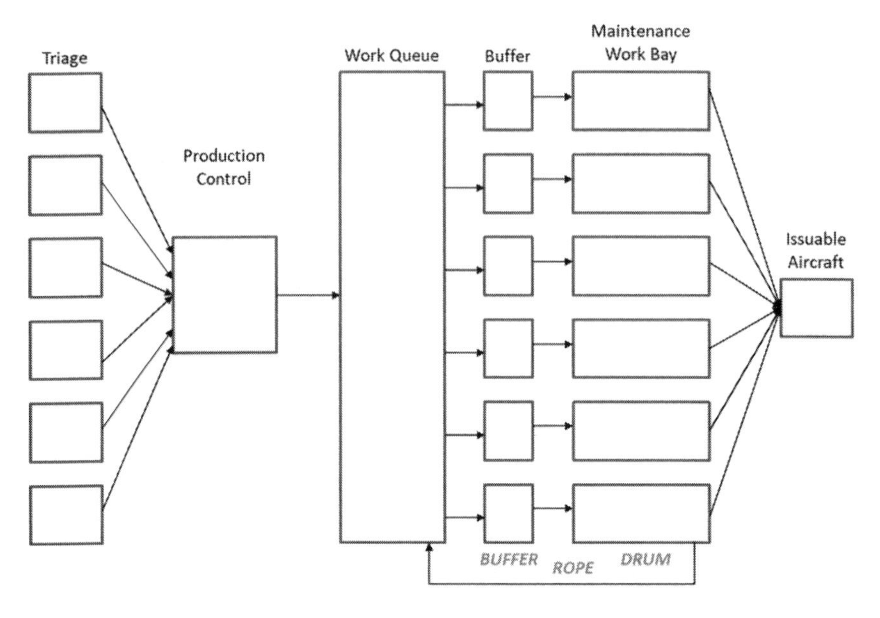

FIGURE 15.8
Multiple Drum Buffer Rope (MDBR).

In the scheduled maintenance area, the implementation of Critical Chain Project Management (CCPM) produced incredible results. On one particular aircraft type prior to this implementation, the average turn-around time was approximately 70 days. After CCPM, the TOC Replenishment System, Full Kitting and the integrated TOC, Lean and Six Sigma training were fully implemented, the turn-around time was reduced to approximately 25 days! In fact, because the turn-around time had improved so much, it was clear that this DoD training facility did not need to have as many aircraft on site. Because of this, this training site was able to return three aircraft to the DoD for assignment to other locations. This action meant that the DoD did not need to purchase three aircraft, thus saving many millions of dollars. And the other good news that occurred because of the significant improvement achieved was that this contractor had their contract renewed. Figure 15.9 is the final version of the Scheduled Maintenance Goal Tree with the injections (i.e. improvement solutions) at the base of the tree.

Because the goal in the scheduled maintenance area still had not yet been achieved, many improvements were achieved. While this case study

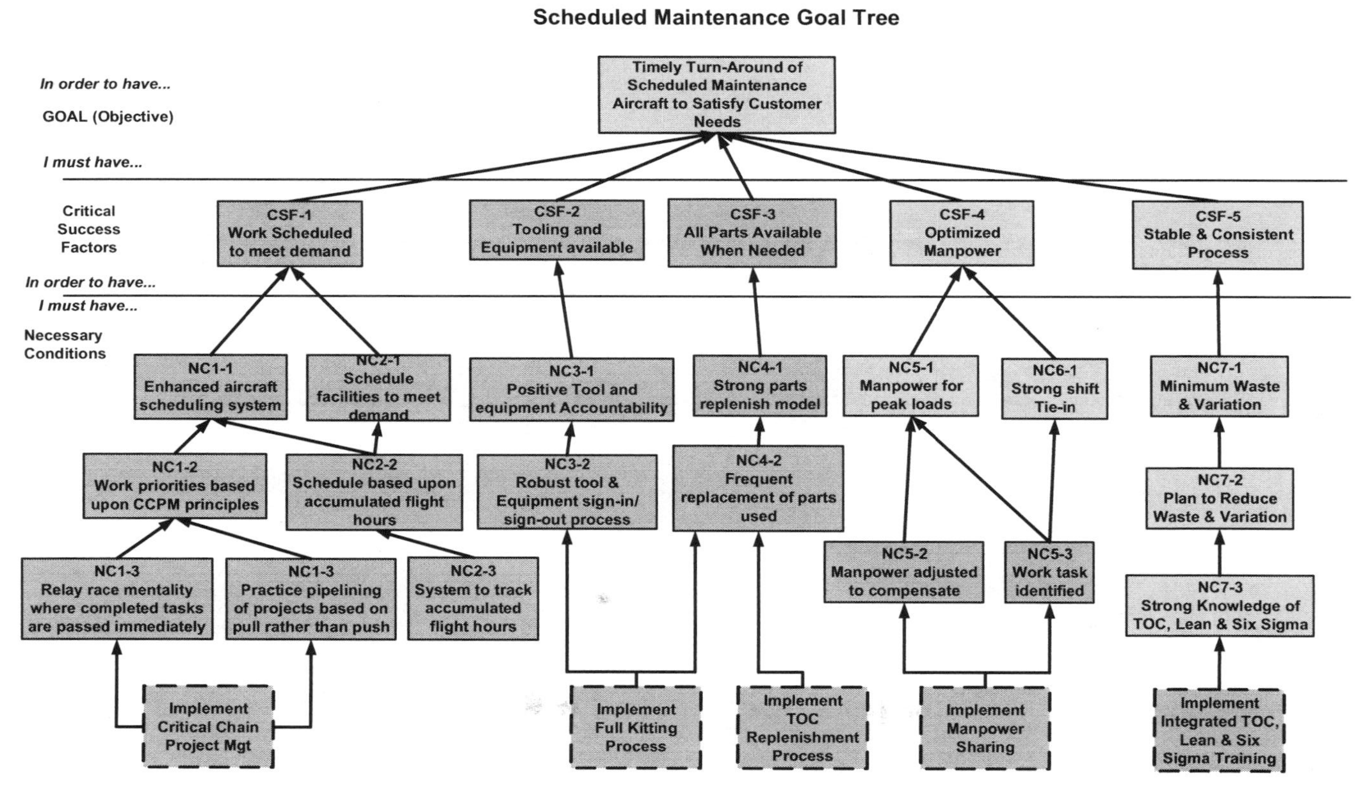

FIGURE 15.9
Final Version of the Scheduled Maintenance Goal Tree with Injections.

had a happy ending, the real difference was that the focus was no longer on cost reduction, but rather it was now on throughput improvement. And the positive results clearly demonstrate why we always focus on increasing throughput and not on cost reduction. Which method appeals more to you?

16

TOC in Healthcare

In this chapter we're going to shift gears from the norm and demonstrate how the continuous improvement methods we have discussed in previous chapters work very well in a medical environment. When you think about it, being in a hospital atmosphere is really not much different than being in a manufacturing setting or even a Maintenance, Repair and Overhaul (MRO), facility as we just discussed in the previous chapter. The patients are the "products" that go through the various hospital processes. The patient enters the hospital, let's say via the emergency room (ER), is then observed by a doctor for symptoms and has tests run to verify what the symptoms might be suggesting; the treatment plan is created and executed, and the patient is either discharged or is admitted to the hospital. The bottom line here is that improvement is the focus.

If you were to create a high-level process map of what takes place in a hospital scenario, it might look something like Figure 16.1.

This scenario is really not much different than what you might see in a manufacturing or MRO setting, so doesn't it make sense that those same tools and techniques for improving the manufacturing or MRO process should work in a hospital setting? Let's now add some estimated times to each of the previous process steps, and see what we see in Figure 16.2.

Let's assume we are interested in speeding up the process for admitting the patient to the hospital, or whether we treat and release the patient from the ER. Which process step is the system constraint, or maybe we should ask it this way: which process step would you focus your improvement effort on to improve throughput through the ER? Clearly, the most significant amount of time in Figure 16.2 would be the time waiting for the test results to come back. This step is necessary so the decision as to whether the patient is treated and released or admitted into the hospital can be made.

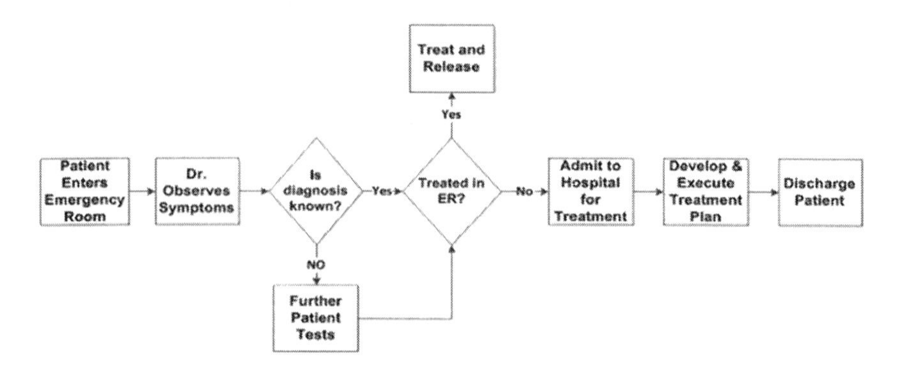

FIGURE 16.1
High-Level Process Map of What Takes Place in a Hospital Scenario.

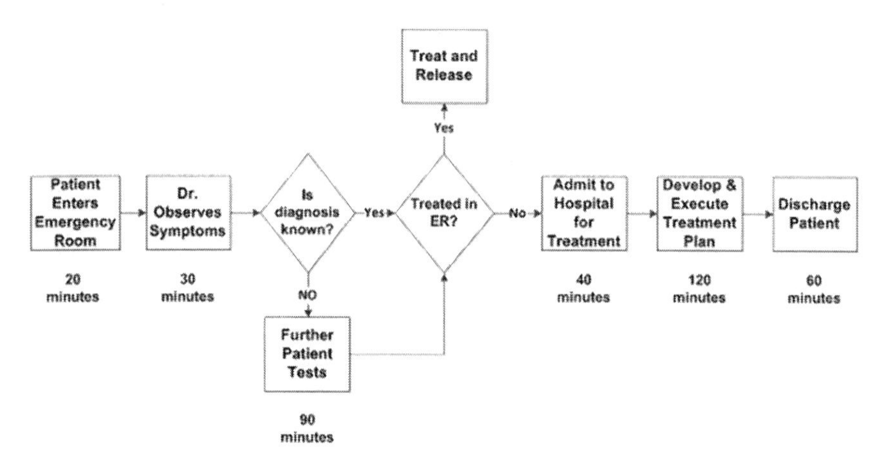

FIGURE 16.2
Process Steps with Times Added.

So, in this very simple example, we see that identifying the system constraint and then focusing our improvement effort there, and only there, yields the greatest opportunity for improvement, even in a hospital environment. Like we've said many times in this book, one of the reasons many Lean Six Sigma initiatives are failing is because there is a seemingly maniacal approach that says fix everything, when in reality finding and focusing on the system constraint is the most effective improvement method.

I've received a lot of emails about whether it's a good idea to use an integrated Lean Six Sigma and Theory of Constraints (or as we also refer to it, Constraints Management) in a healthcare environment. Our response to these questions is no, because it's not a good idea, it's a great idea!!

If we were to ask healthcare professionals if there were flow issues within their hospitals or other healthcare facilities, we doubt seriously that we'd get any answer other than yes. In our experiences, especially during our children's early lives, we've all had to use an emergency room on more than one occasion. And each time we were there, we experienced an enormous amount of idle time just sitting and waiting for our kids to be examined. We all understand that there were probably true emergencies that caused some of this, but even after the initial examination we have all experienced long periods where we just sat and waited. We also remember times when one of our kids had surgery scheduled, and it seemed like we had to wait well beyond the scheduled time before they even moved our child to the surgical unit.

There are drug mix-ups, medicine shortages and a host of other healthcare related issues facing us. Let's face it, the healthcare system in the United States is a mess! So, what does this have to do with using an integrated Lean Six Sigma and Constraints Management, you may be wondering? My answer is everything!

If we continue down the same path we're on right now, we know that our healthcare system will simply get worse. We know that the flow problems, drug mix-ups and shortages of medicine will only intensify, unless we do something drastically different than we are today, and the solution can't be to simply work harder. Healthcare professionals are working as hard as they can already, and many suffer from burn-out caused by working long hours. Lean and Six Sigma will help solve some of the problems, but will they help fast enough before the onslaught of new patients enters the mix? Although both Lean and Six Sigma (or the hybrid Lean Sigma) have their merits, in my opinion they won't get us to where we need to be quick enough or far enough! So, you're probably thinking, OK Bob, then what do you think is the answer to our country's healthcare dilemma?

My answer, at least in the short term, is to radically improve the flow of patients through our hospitals! So how do I propose to do that, you're probably thinking? I will give you the same answer I have to manufacturing companies who were facing significant increases in orders, or to MRO facilities that aren't delivering enough products to their customers and are guilty of late arriving products when they do ship. The answer lies imbedded within the Theory of Constraints (TOC) or Constraints Management, whichever you prefer to call it. Just like the name of this book, it's all about *focus and leverage*. Focus on the system constraint, and leverage the opportunity to reduce the cycle time of the constraint.

We've written about Dr. Goldratt's 5 Focusing Steps, and we have paraphrased his five steps for healthcare settings as follows:

1. *Identify* the system constraint. Why and where are patients getting hung up? Maybe it's an outdated policy or procedure? Maybe it's an actual bottleneck? But you must find it!
2. Decide how to *exploit* the constraint. Use all of your Lean and Six Sigma tools, and focus them on the constraint you've just identified. Maybe your process is totally unsynchronized? Maybe you'll have to use root cause analysis or TOC Thinking Process tools to identify a core problem, but decide what you're going to do.
3. Subordinate everything to the system constraint. It makes no sense at all to push patients into your constraint, because they'll just stack up and waiting rooms become full. Pull your patients through your process at the same rate that the constraint is working, but make the constraint work faster than it is now.
4. If necessary, elevate the system constraint. Before this step you shouldn't have spent any money (or very little), but if you haven't broken the constraint, you might have to spend some. You might need a new test machine, or you might need to hire more cleaners to get rooms ready, but think of the benefits when the constraint moves. It means that you have broken the original one, and your process is moving patients through it at a much higher pace.
5. When the constraint moves, go back to Step 1 and repeat the process. One thing to remember, the constraint will move once you've applied Lean and Six Sigma to it to reduce waste and variation, so get ready to move to the constraint's new home.

What I'm suggesting to you is to use this integrated methodology because it works. Will it completely solve the on-rush of new patients coming down the road? Probably not, but it will at least buy you some time. The key point to remember when you focus your improvement efforts on the system constraint is that you will improve your throughput. That is, in the case of a hospital, a much higher rate of patients passing through the emergency room. A much higher rate of patients passing through surgery. And probably shorter stays in the hospital.

One of the problems facing all healthcare facilities is the area of inventory of medicines, supplies, etc. It seems that hospitals are often running out of an important Stock Keeping Unit (SKU) when it's needed most.

To guard against these SKU stock-outs, often the size of the inventory is increased, and even though there's more inventory in the hospital stock room, shortages still occur. It's frustrating, to say the least! In fact, one of the primary drivers of the constantly increasing cost of healthcare is the cost of materials management. Some studies have estimated this cost could go as high as 50 percent of a typical hospital's budget. Because of this, there is ever-increasing pressure from the hospital's accountants to limit the inventory of SKUs.

One of the Theory of Constraints Thinking Process tools is the Conflict Resolution Diagram (CRD), the purpose of which is to visualize and resolve conflicts. In Figure 16.3 we see such a diagram, which demonstrates the opposing points of view or the conflict in materials management. Both sides have the same objective and want to improve financial results (labeled Objective A). But one side wants to protect the available supply of SKUs by increasing the amount of inventory, while the other side wants to reduce the cost of inventory by reducing the amount of on-hand inventory—and therein lies the conflict or opposing points of view.

The CRD is a necessity-based logic structure, which uses the syntax, *in order to have "x," I must have "y."* From our CRD, one side says, "In order to improve financial results, I must reduce the cost of inventory, and in order to reduce the cost of inventory, I must reduce the amount of on-hand inventory." The other side is saying that "In order to improve financial results, I must protect the supply of SKUs, and in order to protect

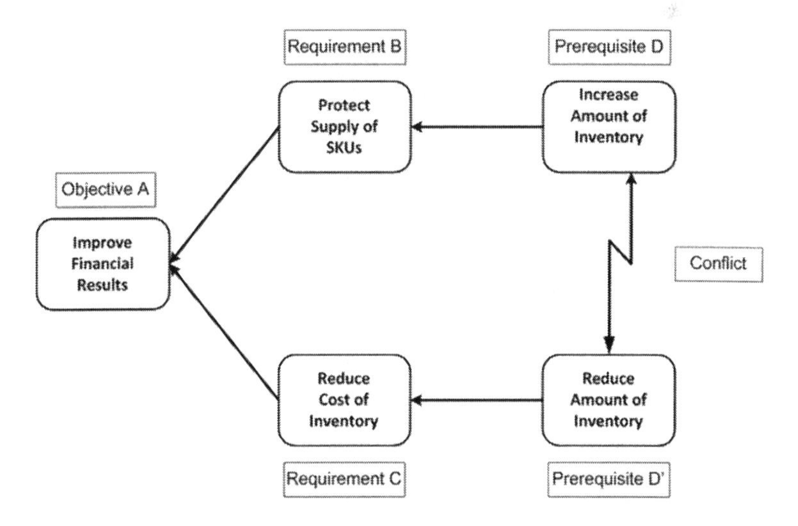

FIGURE 16.3
Conflict Resolution Diagram (CRD).

the supply of SKUs, I must increase the amount of inventory." Clearly there is a conflict between the two points of view. So how do we break this conflict?

One of the teachings of TOC is that compromise solutions should never be considered, because one side typically gets what they want while the other side loses, or both sides get *some* of what they want. What we should be striving for always is a win-win solution!! An optimum solution, if you will. The way we do this is to surface the assumptions behind each side's assertions. The assumptions are the *because* statements that explain each point of view. For example, one side is saying "In order to improve financial results, I must reduce the cost of inventory" *because* excess inventory hurts our cash flow position. Maybe the other side is saying, "In order to improve financial results, I must protect the supply of SKUs" *because* if I run out of an SKU it could decrease Throughput. If we can insert an *injection* or solution that counters both sets of the "*becauses*," then we have resolved the conflict. So, with this said, the solution must both reduce inventory while protecting throughput. Sound like fun?

Before we have fun resolving this conflict, let's go into a bit more depth on why what a typical healthcare facility is using today isn't working so well. Most, if not all, businesses are linked one way or another to some kind of supply chain. They need SKUs or raw materials from somebody else to do what they do and pass it on to the next system in line until it finally arrives at the end consumer.

In a hospital environment, this means that a doctor will prescribe a treatment that requires something to bring the patient back to a healthy state. Maybe it's a prescription for a medicine or maybe even an orthopedic item, so he writes the prescription and waits for the order to be filled and then given to his patient. Hopefully what the doctor has prescribed is in the stockroom, so the patient can begin the treatment.

For many organizations the supply-chain/inventory system of choice is one often referred to as the Minimum/Maximum (Min/Max) system. In this type of supply-chain system, SKUs (or inventory) are evaluated based on a projected need and usage (a forecast), and some type of maximum and minimum levels are established for each item. The typical rules that are followed for these Min/Max systems are

Rule 1: Determine the maximum and minimum levels for each item.
Rule 2: Don't exceed the maximum level.
Rule 3: Don't re-order until you go below the minimum level.

The foundational assumptions behind these rules and measures are primarily based on the belief that in *order to save money and minimize your expenditures* for supply inventory, you must minimize the amount of money you spend for these items. Remember Figure 16.2, the Conflict Resolution Diagram? The assumption here is that the purchase price per SKU (unit) could be driven to the lowest possible level by buying in bulk, and the company would *save* the maximum amount of money on their purchase. The reality is that there always seem to be situations of excess inventory for some items and of stock-out situations for others. So why is it that even though we have plenty of inventory, these stock-outs continue to happen? Let's take a more in-depth look at the typical rules for managing this Min/Max Supply System.

1. The *system re-order amount* is usually the maximum amount, no matter how many SKUs are currently in the point-of-use (POU) storage bin. The thinking here is that to obtain the maximum discount, we must always buy in bulk.
2. Most supply systems only allow for one order at a time to be present in the system for a specific SKU.
3. Orders for SKUs are triggered only after the *minimum amount has been exceeded*. That is, for example, if the minimum level for a drug is set at 1000, when it goes below 1000 it can be reordered.
4. Total SKU inventory is held at the lowest possible level of the distribution chain, the POU storage location. Typically, this is at the hospital unit or ward.
5. SKUs are inventoried once or twice a month and orders placed, as required.

Graphically, Rules 1, 2 and 3 look like what you see in Figure 16.4. The problem with this system is that it's prone to conditions of stock-outs on a fairly routine basis, as depicted in Figure 16.5 where the pattern repeats itself.

In Figure 16.5 we see that even though there was inventory in the system, we still have the stock-out situation happening. Rules 4 and 5 are graphically illustrated in Figure 16.6. In Figure 16.6 we see that parts are distributed from the supplier and *pushed* down through the links in the supply chain, which ends up clogging the supply chain with disorganization and ineffectiveness.

The classic symptoms that we see using the Min/Max system are

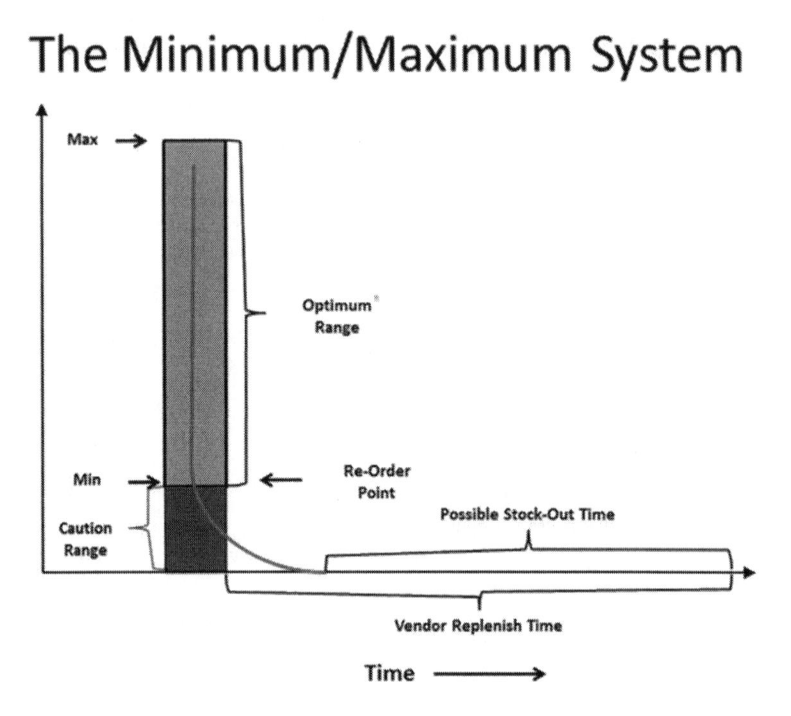

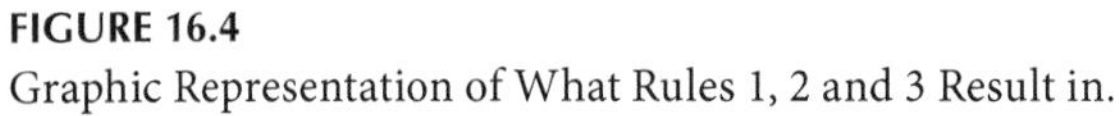
FIGURE 16.4
Graphic Representation of What Rules 1, 2 and 3 Result in.

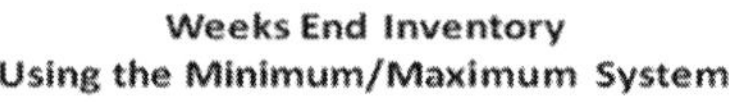

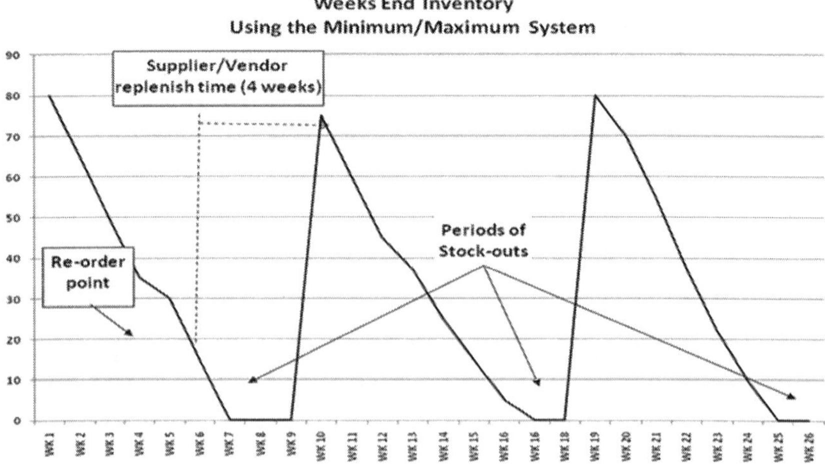

FIGURE 16.5
Impact of Min/Max over Time.

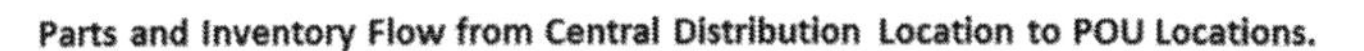

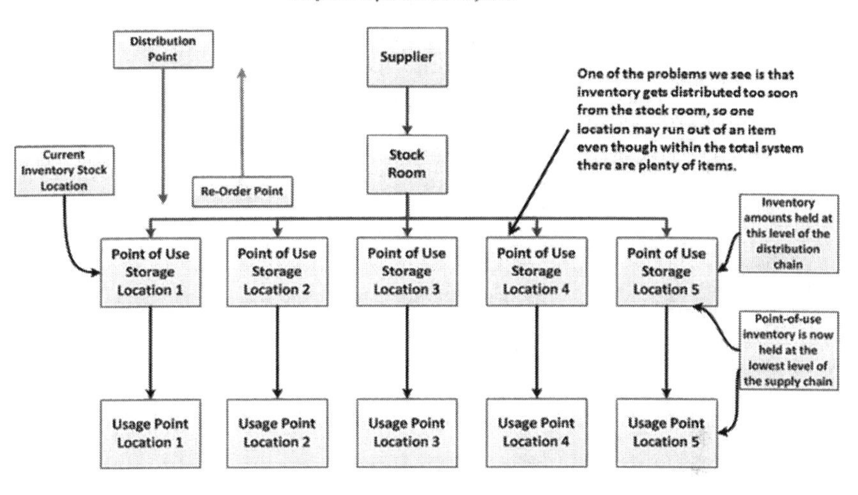

FIGURE 16.6
Depiction of Parts Distributed from the Supplier and Pushed down through the Links in the Supply Chain.

- Having the wrong medicines or drugs in stock or having too much of one and too little of another.
- Significant amounts of cash tied up in excessive inventory.
- Significant amounts of money lost through drug obsolescence or even being out of date.
- Hoarding of SKUs by nurses and other medical personnel so that they have their own personal stash when needed. They all mean well because they have the best interests of the patient in mind.

So whatever system we are proposing here, it must overcome these negatives in a significant way. Before we explain how TOC's Replenishment System works, let's look at some of the basic differences between it and the Min/Max system. First of all, the TOC Replenishment Model (TRM) is a robust SKU replenishment system that allows the user to be proactive in managing the supply-chain system. And unlike the Min/Max system, which *pushes* material down the supply chain based upon a minimum reorder inventory target, TOC's model is a *pull-based* system, whereby material is replenished based upon actual usage. This is, in my mind, the key differentiator between the two replenishment methods.

The TOC Replenishment Model (TRM) argues that the majority of the inventory should be held at a *highest* level in the distribution system (supply chain) and not at the *lowest* level as mandated by the Min/Max model. This is an important difference, because often materials get distributed too early to users resulting in stock-outs in one part of the supply chain and potentially excess inventory in another.

Unlike the Min/Max system, instead of using the minimum amount in inventory to trigger the re-order process, the TOC Replenishment Model is triggered by daily usage and supplier lead times to replenish. This quality, plus the location of the inventory stock, virtually eliminates the aforementioned stock-outs in parts of the supply chain. The TRM eliminates the maximum re-order amount as well and replaces it with an increase in the order frequency, again based upon usage. This change in order frequency based on usage effectively reduces the volume of on-hand inventory, typically on the order of 30 percent to 50 percent!

Ok, so now that we have seen some of the key differences between the Min/Max system and the TRM system, let's now focus on how the TRM system overcomes the negative symptoms of the Min/Max system.

As stated earlier, in the TRM system, stock is positioned at the highest level in the distribution system so that all available inventory can be used to satisfy demand at the downstream multiple points of use. Since the location of the stock is positioned in this way, this allows more frequent ordering to be completed. The central warehouse or stock room sums the demand usage of the various usage points so that larger order quantities can be accumulated at the central stock location sooner than at each separate location.

In the TRM, *buffers* are positioned at points of potential high demand variation and stocked and restocked at levels determined by stock on hand, demand rate and replenishment lead time. Order frequency is increased and order quantity is decreased to maintain buffers at optimum levels and, as a result, stock-out conditions that cause interruption to the flow of patients are usually completely avoided.

Rather than relying on some minimum stock level to trigger a re-order of SKUs, ordering is determined by the depletion of the buffer stock. So effectively, how much to order and where to distribute the available stock is determined by the status of the buffer for that SKU. The data for depletion of buffer provides signals to determine when and how much to modify buffer size.

Order urgency is based on the depletion of the buffer and is therefore used to set ordering priorities. The TRM order method accounts for buffer depletion and local demand information so the right mix of SKUs is ordered, and SKUs are distributed to the priority locations. This is one of the major differences between the two replenishment methods.

There are key criteria that must be in place for the TOC Dynamic Replenishment Model to work effectively, as follows:

1. The system re-order amount needs to be based on daily or weekly usage and SKU lead time to replenish.
2. The system needs to allow for multiple replenish orders, if required.
3. Orders are triggered based on buffer requirements, with possible daily actions, as required.
4. All SKUs/inventory must be available when needed.
5. SKU inventory is held at a higher level, preferably at central supply locations or coming directly from the supplier/vendor.
6. SKU buffer determined by usage rate and replenish supplier lead time. Baseline buffer should be equal to 1.5. If lead time is one week, buffer is set at 1.5 weeks, and then we can adjust the size as required, based on historical data.

The TOC Distribution and Replenishment Model tells us that we should hold most of the inventory at the highest level in our supply chain and not at the lowest level like the Min/Max system. Yes, we still want inventory at our point of use, but not the majority of it. One of the major consequences of the Min/Max system is the distribution of SKUs much too early, especially when the same type of inventory or part is used in several locations such as different hospital wards or units. It's not uncommon to see, for the same SKU, an excess in one ward and a stock-out in another, all because the inventory was *pushed* down through the supply chain. This does not happen in the TOC Replenishment Model since stocks are *pulled* through the system based upon usage.

In the TRM we eliminate using the minimum target as a trigger to re-order and replace it with a system that monitors our safety buffer and usage on a daily or weekly basis, and replenish only what has been used for that time period. We also eliminate the Min/Max maximum order quantity in that we only order what has been *consumed* rather than some maximum level. What we end up with by using TRM is much lower inventory levels, in the right location, at the right time, with zero or minimal stock-outs. In fact, using TRM we not only virtually eliminate stock-outs, but we do so usually with 40 percent to 50 percent less inventory, thus freeing up huge amounts of cash.

To further make this point, I want to use a very common example taken from a book I co-authored with Bruce Nelson, *Epiphanized*, in Appendix 5. In this scenario, we tell readers to consider a soda vending machine. When the supplier (the soda vendor) opens the door on a vending machine, it is very easy to calculate the distribution of products sold, or the point-of-use consumption. The soda person knows immediately which inventory has to be replaced and to what level to replace it. The soda person is holding the inventory at the next highest level, which is on the soda truck, so it's easy to make the required distribution when needed. He doesn't leave six cases of soda when only 20 cans are needed. If he were to do that, when he got to the next vending machine he might have run out of the necessary soda because he made *distribution too early* at the last stop.

After completing the required daily distribution to the vending machines, the soda person returns to the warehouse or distribution point to replenish the supply on the soda truck and get ready for the next day's distribution. When the warehouse makes distribution to the soda truck, they move up one level in the chain and replenish from their supplier. This type of system does require *discipline* to gain the most benefits. It assumes that regular and needed checks are taking place at the inventory locations to determine the replenishment needs. If these points are not checked on a regular basis, it is possible for the system to experience stock-out situations.

Remember how we demonstrated the effects of the Min/Max replenishment method with what you see in Figure 16.5. What you see in Figure 16.5 are the results of a simulation we ran using the following criteria (For details and actual data used please refer to Appendix 5 in *Epiphanized*):

1. The maximum level is 90 items.
2. The minimum re-order point is 20 items.
3. The lead time to replenish this SKU from the vendor averages four weeks. The average is based on the fact that there are times when this SKU can be delivered faster (three weeks) and other times it delivers slower (five weeks).
4. Usage of this SKU varies by week, but on average is equal to about ten items per week.

Remember, when using the Min/Max replenishment method we don't re-order until we meet or exceed our minimum re-order quantity (i.e. 20 items left in stock). In Figure 16.7 we are applying the TOC Replenishment

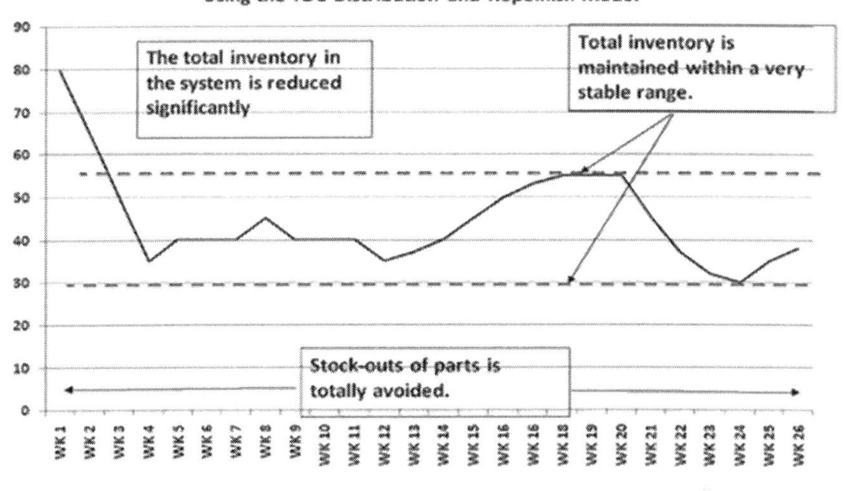

FIGURE 16.7
Results of Applying the TOC Replenishment Model.

Model rules to exactly the same criteria we set for Figure 16.5. We used the same SKU simulation and the same period of time, with the same usage numbers. The difference will be in this simulation we changed the rules to fit the TOC Replenishment Model. That is, re-order is based on usage amount and vendor lead time rather than minimum and maximum amount.

In Figure 16.7 we assumed the following:

1. Maximum level is 90 items. (This is the start point for the current inventory when the TOC Replenishment Model was initiated.)
2. There is *no minimum re-order point.* Instead, re-order is based on *usage* and *vendor lead time.*
3. Lead time to replenish is still four weeks.
4. Average usage of the part is about ten per week.

There are several key points observed in Figure 16.7:

- What is most notable is that total inventory required through time has been virtually cut in half when compared to that of Figure 16.5.
- There are no stock-out situations present.
- The total inventory is maintained within a very stable range over a long period of time.

Searching for SKUs and having to experience the negative impact of stock-outs are constant problems in many hospital supply-chain systems. These problems aren't caused by the logistics people, but are instead the negative consequences of the supply-chain system and the manner by which it is used. The concepts and methods associated with the TOC Replenishment Model can and will positively impact the flow and availability of SKUs within a hospital setting.

17

Healthcare Case Study

In this chapter we're going discuss a case study from the healthcare field. We will discuss a recently completed Process Value Stream Analysis (PVSA) project at a hospital located in the Mid-Western part of the United States. The focus of this PVSA was in this facility's Emergency and Cardiology Departments, where they wanted to improve one of their key performance metrics, Door to Balloon (D2B) time. For those of you (like me before I started this engagement) who don't have a clue as to what D2B is, let me fill you in. Door to Balloon is a time measurement in Emergency Cardiac Care (ECC), specifically in the treatment of ST Segment Elevation Myocardial Infarction (or STEMI heart attack).

The interval starts with the patient's arrival in the Emergency Department and ends when a catheter guide-wire crosses the culprit lesion in the Cardiac Cath lab. In everyday language, this just means that a balloon is inflated inside one of the heart's primary blood vessels to allow unimpeded blood flow through the heart. The clock starts ticking either as a walk-in to the Emergency Department or in the field where a patient is being attended to by medical personnel. This metric is enormously important to patients simply because the longer this procedure is delayed, the more damage occurs to the heart muscle due to a lack of oxygen to the heart muscle. It's damaged because the cause of this problem is typically due to a blockage within the heart that prevents oxygen from being supplied to the heart, and without proper amounts of oxygen, muscle damage results. The inflated balloon "unclogs" the blood vessel.

I started this event with a training session for the team members focusing on how to use an integrated Lean Six Sigma and Constraints Management. We have seen a lot of PVSAs where waste is identified throughout the process, and then the team works to either reduce it or eliminate all of it. It has been our experience that when attempting to reduce the time it takes to process something through a process such as this one, by attacking the entire process for waste reduction, teams frequently miss the opportunity to reduce the cycle time much more quickly than they otherwise could have. This is where the Theory of Constraints (TOC) and its 5 Focusing Steps offers a much quicker solution to this type of project. Just to review, TOC's 5 Focusing Steps, first introduced by the late Dr. Eli Goldratt, are

1. *Identify* the system constraint—In a physical process with numerous processing steps, the constraint is the step with the smallest amount of capacity. Or another way of stating this is the step with the longest processing time.
2. Decide how to *exploit* the system constraint—Once the constraint has been identified, this step instructs you to focus your efforts on it and use improvement tools of Lean and Six Sigma to reduce waste and variation, but focus your efforts mostly on the constraint. This does not mean that you can ignore non-constraints, but your primary focus should be on the constraint.
3. *Subordinate* everything else to the constraint—In layman's terms this simply means don't over-produce on non-constraints, and never let the constraint be starved. In a process like the Door to Balloon time, it would make no sense to push patients into this process, since they would be forced to wait excessively. But of course, the hospital cannot predict when patients with heart attacks will show up needing medical attention. But by constantly trying to reduce the constraint's time, the wait time should be continuously reduced.
4. If necessary, *elevate* the constraint—This simply means that if you have done everything you can to increase the capacity of the constraint in Step 2 and it's still not enough to satisfy the demand placed on it, then you might have to spend money by hiring additional people, purchasing additional equipment, etc.
5. *Return to Step 1*, but don't let *inertia* create a new constraint—Once the constraint's required capacity has been achieved, the system constraint could move to a new location within the process. When this happens, it is necessary to move your improvement efforts to the new constraint if further improvement is needed. What is thing about inertia? What Goldratt meant by that was to make sure things you have put in place to break the original constraint (e.g. procedures, policies, etc.) are not limiting the throughput of the process. If necessary, you may need to remove them.

For whatever reason, the agency who developed this universal metric used the *median* rather than the mean. The current median standard for Door to Balloon time was set at 90 minutes, and this hospital was actually doing quite well against this standard with a median score of 66 minutes. However, because this hospital is anticipating the standard would be changing to 60 minutes in the future, they decided to be proactive by putting together a team of subject matter experts to look for ways to achieve this future target before it is mandated to do so. In addition to this new time benefiting the patient (i.e. much less heart muscle damage), there is also a financial incentive for the hospital in that reimbursement rates for Medicare and Medicaid patients are tied to completing the D2B time below the standard median time.

After completing the training session, the team was instructed to "Walk the Gemba" by going to both the Emergency Department and Cardiology to observe what happens during this process and to have conversations with employees from both departments about problems they might encounter. This was a fact-finding mission aimed at understanding how patients are managed through this treatment process. The team collected many observations during this walk, most of which would be used to construct their Current State Process Map, which unfortunately is too large to post here. The team also had access to D2B time data that has been collected on previous patients passing through this process. The team then analyzed the data to better understand what was happening on previous D2B events. Figure 17.1 is a summary of this analysis.

The time data that had been collected was broken down into three separate phases of the D2B process: Door to EKG, EKG to Table and Table to Balloon. This was extremely helpful for the team in their efforts to identify the system constraint. As you can see in Figure 17.1, the EKG to Table Phase, with a mean value of 36.7 minutes, is clearly that part of the process requiring the most time and was designated by the team to be the system constraint. Table to Balloon time, at 21.2 minutes on average, also consumed a significant amount of time, while Door to EKG only required 4.75 minutes to complete. It is important to remember that this metric (D2B) was developed to capture median times rather than mean times, so hospitals are judged (and reimbursed) by a median time and are reported as such. The difference between the median and mean times for EKG to Table (i.e. median = 32 minutes and mean = 36.7 minutes) indicates that the data might be skewed and not perfectly normally distributed. This, of course, means that there are outliers that must be investigated for cause.

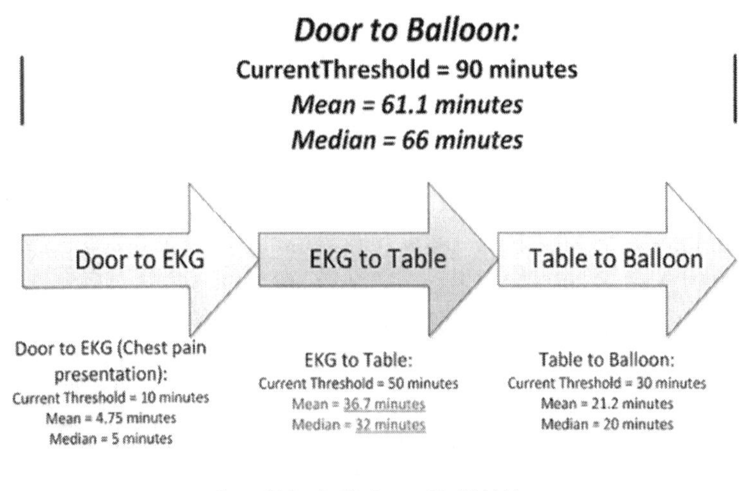

FIGURE 17.1
Summary of D2B Analysis.

After collecting and analyzing this data, the team was instructed to create two Interference Diagrams (IDs), one for Phase 2 (EKG to Table) and one for Phase 3 (Table to Balloon). You may recall from an earlier chapter that the purpose of the ID is to identify any barriers or obstacles (i.e. interferences) that stand in the way of achieving a goal or objective. In the cases for Phase 2 and 3, the goal was identified as reducing the time required to complete each phase. Figure 17.2 was of the ID created for the EKG to Table phase and is presented here only to depict what an Interference Diagram looks like for those of you who may never have used one before. The post-it notes contain a description of the interference with an estimate of how much time the interference might negatively impact the goal of reducing cycle time.

The team then used their fact-finding "walks" (i.e. observations and conversations) and the Interference Diagrams to create both an Ideal State and a Future State Map. The next image compares the Current State Map and Future State Map after completion of the standard value analysis, and as can be seen, the number of total steps was dramatically reduced, as were the number of decision points, swim lanes and hand-offs. You will also notice, when comparing the current state to the future state, a dramatic reduction in the number of non-value-added (i.e. red) steps (i.e. 27 to 2). The team also developed an Ideal State Map, which is included in Figure 17.3.

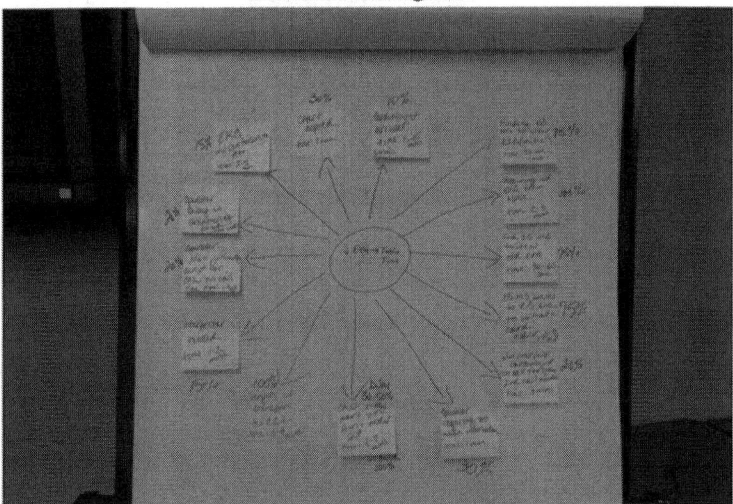

FIGURE 17.2
Photo of ID Created for the EKG to Table Phase.

This team did an excellent job of analyzing this important process and was able to remove much of the waste contained within it. But the real improvement came in the overall potential time to complete this procedure, which should have a significantly positive impact on damage to patient's heart muscles when their recommendations are implemented, and this

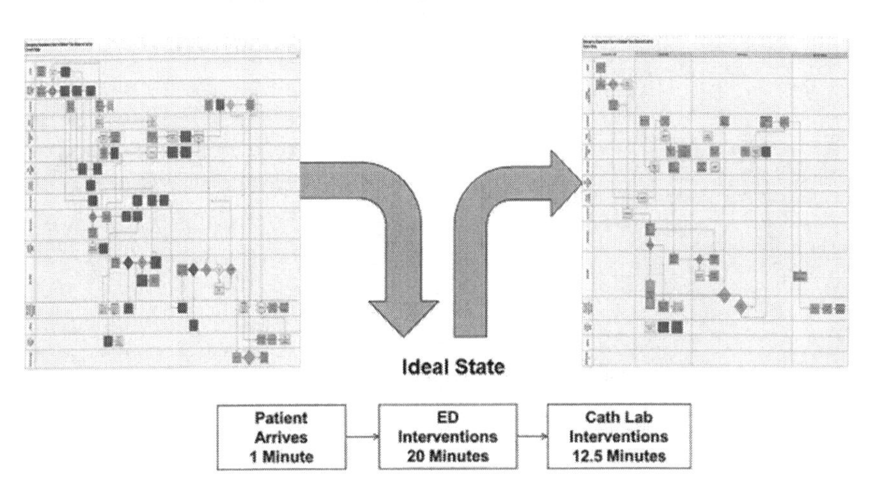

FIGURE 17.3
Current State, Ideal State and Future State of D2B Time.

Metric	Pre-Event	Post-Event	Improvement
Total Number of Steps	81	49	-32
% of Value-Added Steps	37 %	61 %	24 %
# of Swim Lanes	19	14	-5
Cycle Time	66 Min	~53 Min	~ -13 Min
# of Decisions	13	6	-7
# of Green Steps	26 (3)	29 (3)	+3 (0)
# of Yellow Steps	16 (3)	10 (4)	-6 (+1)
# of Red Steps	27 (6)	3 (0)	-24 (-6)

FIGURE 17.4
Actual Cycle Time Reduction Achieved by Team.

was the overriding premise and objective of this event. The following is a summary of before and after for this PVSA. The highlighted numbers in Figure 17.4 represents the actual cycle time reduction. Although a reduction in Door to Balloon time of 13 minutes might not seem like much of an improvement to some of you, consider how much less damage to a patient's heart muscle might be avoided. In the healthcare field for procedures such as Door to Balloon time, every minute counts.

18

The Cabinet Maker

THE CONSEQUENCES AND CHAOS OF UNCONTROLLED EFFICIENCY

The ultimate story of the Cabinet Maker could actually be any company, in any industry, in any location, and at any point in time. However, for this story we will focus on a particular cabinet maker and their improvement journey.

This cabinet maker, by industry standard, could have been classified as a small company. It had grown from a father-and-son operation, conducted in a backyard woodshop (shed), into a company that had annual revenues of about $1.5M and employed about 30–40 people at any given time. In time, they moved from the backyard woodshop into a sizable building complete with its own shipping and receiving docks, plenty of floor space for required equipment and a large area to hold raw material inventory—complete with storage racks that went all the way to the ceiling, probably about 30–35 ft. high. The new building also contained adequate area to lay out and build special cabinets, such as for reception areas. There was also an adequate area where counter tops could be manufactured.

This cabinet maker was considered specialized because the highest percentage of their work was commercial cabinetry and not residential cabinetry. The primary business model focused on office buildings, schools and other larger public structures (e.g. libraries) that required cabinets and sometimes the very special (very ornate) reception desks located in office lobbies and waiting areas. This cabinet maker could also provide a wide range of custom counter tops, if required. In essence, they were very much a "one-stop shop" to provide the cabinets, custom cabinets, custom counter tops and special design cabinets for larger projects.

For the most part, the cabinet maker had limited their work efforts within a single state, but they had recently made a management decision to expand the sales effort into three adjoining states. Management was of the mindset that they needed to grow the business. By expanding the sales into additional states, they were able to increase the sales and had done quite well getting new business. Actually the "sales" statement is a bit of a misnomer. What they actually did was expand their effort to provide "bids" to local contractors that were bidding and working projects for local, county and state governments, and on several occasions they bid and won contracts for the federal government, such as court houses. This expanded bidding process seemed to be working well, and the orders were rolling in. They were winning some major subcontracts to provide cabinetry in several new schools that were being built—life was good! Except for one small thing—the new projects were coming in fast, but they weren't going out fast enough to support all of the new orders!

The new projects seemed to be *stalled* in the production process. It was not uncommon that new projects could take as long as two months to complete, and this was not a good situation. The cabinet maker couldn't get paid for projects that were not complete, and because of some contract limitations (which seemed to be typical for this industry), they could not receive progress payments. Cash flow was a major concern, and rightly so. In fact, in some cases the cabinet maker was being charged a severe penalty by the prime contractors for late deliveries.

For a majority of the new business, the cabinet maker was a subcontractor to several different major construction companies who were the primary builders for the projects. As such, the primary contractor used MS Project to schedule their project(s). MS Project is a Critical Path Method (CPM) scheduling system, and that by itself presented a multitude of recurring issues. The cabinet installation was typically one of the last scheduled task segments to be completed, and as such, they were always near the end of the project schedule. This problem usually manifested itself when any task slack time in the previous tasks had already been consumed by previous tradesmen. The cabinet maker was often exposed to a reduced time-setting to complete their tasks and was often asked to complete tasks in less time than they had originally bid.

To compound the problem even further, if the cabinet maker did not perform the tasks in the reduced time offering, they could (most of the time) be subjected to a very severe financial penalty for not completing the work on time. The construction companies certainly held the high ground and contractual advantage to make this a very painful process. It was a vicious cycle of reduced cash flow, poor on-time delivery and the inability to accept any new work. This finally pushed the cabinet maker management team to ask for some help.

THE PRELIMINARY ANALYSIS AND INFORMATION GATHERING

When the improvement team arrived, they met with the management team to get a clear understanding of the current issues, which was basically a repeat of what has previously been stated with some additional highlights. For starters, the management team was deeply committed to traditional Cost Accounting and efficiency metrics. They supported this thinking with the rationalization that the primary key to their past success was being able to reduce the cost per part (cabinet boxes). Most, if not all, of their decisions were focused on how to reduce the cost per cabinet. It was a continuation of the management method they had learned years before and the management method they brought with them to the present. However, what had seemed to work in the past was now seemingly their worst enemy. In their minds, they were following the "rules," and yet the ship was sinking. It was a continuing cycle of chaos, and they literally had no idea how to break the cycle and get out of the current situation. The only answer they had was to keep doing the same thing and hoping it would work for them.

Management also confessed that they were paying the late fees and having the figure deducted from what was owed to them. They did this in hopes of trying to generate enough cash to have enough money to pay for the raw materials for the next project. However, the raw material situation had escalated to the point that suppliers were starting to refuse orders from them until previous bills were paid in full, or at least paid down to a reasonable level. They concluded that there was no way out except to get their cost per cabinet down to a lower rate, the assumption (and false hope) being that any money they could generate through savings could be applied to other needs. They were convinced (hoped) that if they could increase the efficiency of production, and achieve the lower cost per part, then the current problems would "magically" correct themselves. At this point, the primary goals of the management team were to focus on improving efficiency and reducing costs.

Having gathered the information from the management team interviews, the improvement team shifted their focus to the production system. First and foremost was the visual analysis aimed at documenting the product flow to gain a basic understanding of the rhythm of how the work moved through the system from the start to the finish.

THE PRODUCTION PROCESS AND TOUR

Step 1—The Planners

When a new project was awarded, and the architectural drawings were received, the planners would convert the architectural drawings into shop drawings, including all dimensions, cut sizes and material requirements. This information was used by various workstations to produce the cabinets, drawers, doors and counter tops. The required material list of hardware for drawer pulls, drawer guides, hinges and laminate colors was also provided. If a special cabinet was required, they used the architectural drawing without conversion.

For the most part, the raw material was standardized using 4 × 8 sheets of medium-density fiberboard (MDF), or sometimes plywood. When the shop drawings were completed, the information was loaded into a software program that would optimize the panel layout and cutting sequence. This software was used to determine the maximum cut from a panel to reduce the amount of scrap. This software had been previously purchased as another "Cost Savings" idea by the management team.

Before the job was released to the production floor, it was assigned an internal four-digit project number. The project number became the formal identity for the project while it was in production. The planning group would print the various components of information into a production package with the required information for each work center.

When the job was released, it was also assigned to an internal project manager. The project manager was responsible for making sure the job was continually moving through the production process. This cadre of project managers was usually the more senior personnel who had some experience in all of the workstations and could be called upon for technical assistance when issues occurred.

Step 2—The Lay-up

The laminate lay-up process was performed in tandem with the saw operation. During the lay-up process the raw 4×8 panels, of either MDF or plywood, were fitted with a sheet of laminate to match the color requirement in the production instructions. These raw 4×8 wood panels were sprayed with glue and the laminate applied and rolled into place. The laminate raw material was also purchased in 4×8 sheets. These matching material sizes usually allowed the lay-up process to be completed in short order.

After the panels had been fitted with the correct laminate color they were moved to the saw. This material transfer was usually completed by hand and typically moved one panel at a time. The panels were placed in the saw stacking area. If you think of a clip for bullets, you get the idea of the stacking area. Moving panels from the stacking area to the saw was an automated feeding system that moved the panels, one at a time, onto the saw cutting deck.

Step 3—The Saw

This wasn't an ordinary saw, but rather a very large and complex saw. It was computerized and controlled by a single operator. The computer on the saw was linked to the planner's computer. This link allowed for the transfer of the panel optimizing software from the planners to the saw.

As each panel was removed from the stacking area, and prepared for cutting, the operator would program the saw with the downloaded cutting instructions from the optimization software. The saw push arm would move the panel into place and lock it down. The saw mechanism was actually the opposite of what you might think. In a typical table saw, the saw blade remains stationary while the wood is pushed through to make the cut. In this case, the wood was held in place on the cutting deck and the saw blade moved back and forth. When panels were placed on the cutting deck, the saw would make the cuts according to the programed instructions. When the panel had been cut in one direction, the cut pieces were moved (re-aligned) by the operator to make the cuts in the opposite direction, to end up with the correctly sized and cut pieces.

After cutting, the pieces were stacked on a pallet. Theses pallets were loaded and moved according to a semi-specific batch size, which was about 30 pieces. It was normally the saw operator, using a floor truck, who moved the pallets to the next workstation—Edge Banding.

Step 4—Edge Banding

As the name implies, edge banding was the machine that applied the edging material to the panels. The edging material was usually made from plastic, but could sometimes be made from wood. The panels were processed through the edge banding machine, and the edge material was glued (banded) in place. As the edging material was applied, it went through a short series of pressure rollers to set the glue. The edging also passed over a router to smooth both edges and square the corners. The doors and drawer fronts required four passes through the machine to edge all four sides.

At this point, the piece was considered ready for the next process step, and the panels could move in two different directions. The drawer fronts were moved to the drawer assembly area and were attached to the drawer boxes and moved to final assembly. The doors and box pieces were re-loaded on separate pallets and moved with the same transfer batch in mind to the next workstation—the Morbidelli.

Step 5—The Morbidelli

The Morbidelli, as it was called, was actually a machine brand name rather than a machine function. This was a unique and specialized piece of equipment with a very specific job. The Morbidelli was a drilling machine, with the primary function of drilling the necessary holes into the panels. These holes included the dowel holes on the end and sides of the panels, and the series of holes drilled on the flat surface that allowed for the shelf pins to be inserted. These drilled holes allowed for the shelves to move up and down to various heights.

The drilling function allowed for multiple holes to be drilled at the same time. This Morbidelli was also computer controlled with panel dimensions programed in for accurate placement of the holes. These programed commands were not linked to the optimization software, but rather manually input based on project requirements. When the holes were complete, the panels were moved back to the transfer pallet for movement to the next workstation—the Box Press.

Step 6—The Box Press

The box press was a low pressure hydraulic press that would apply pressure from three sides (top, left and right sides). The bottom was stationary on the press deck. The pressure allowed the glue to move and then set within the dowel holes

When the box pieces arrived, they were laid out on the box press deck for assembly. The assembly included all of the pieces to make a box (both sides, top, bottom, back and any internal supports for drawers). The process was to use small amounts of glue in the dowel holes and insert the wooden dowels and assemble the box.

The back was free-floating to allow for expansion and contraction of the wood. The backs were cut in a different location (not the main saw) and were considered a "feeding" part for box assembly. Each piece marked with the job number and the specific box number it was assigned to. When the box was roughly assembled, a rubber mallet was used to position the dowel pins and bring the box together in a rough snug fit. The box was then placed in the box press where pressure was applied to seal (press) everything together. The box press duration was about three minutes per box. This was amount of time considered necessary for the glue to dry and allow the box to stay together. When the boxes left the press, they moved one box at a time to the next workstation—Final Assembly.

Step 7—Final Assembly

During final assembly the boxes were fitted with the correct doors, which had already been cut and edge banded and were waiting (hopefully) in final assembly. The correct hardware for the hinges and drawer pulls was identified from the production package and added.

The final assembly process was helped by the fact that the holes for door hinges and drawer guides had already been drilled by the Morbidelli. It was simply a matter of attaching the correct hardware and hanging the doors and inserting the drawers onto the drawer guides.

The doors and drawers were marked with job numbers and cabinet numbers to make sure the right doors were put on the correct cabinets, which didn't always work out. Final assembly had numerous repairs and rework because sometimes the wrong doors had been installed on the wrong cabinet. The drawers were assembled in another location and delivered to final assembly. The drawers were marked with the job number and cabinet number to aid the assembly process. This process presented its own form of issues.

When the assembled boxes were complete with doors, drawers and hardware, they moved from final assembly as single units to the next workstation—Shipping.

Step 8—Shipping

At the shipping location, each box was wrapped with plastic wrap to protect the outside laminate from scratches and cushion the stacking process when the boxes were loaded in the truck for transportation. The boxes were checked by project number to make sure they were heading to the right job site. Additional materials, such as base boards, flashing panels and counter tops, were also added to the truck, as required.

There were several instances when the shipping inventory was stacked in several holding areas waiting for a truck. With one truck available, the shipping folks had to wait for the truck to return before loading it again. Often, they had to wait a considerable time for the truck to return, especially when they were delivering to several states! Figure 18.1 is a visual representation of the cabinet flow.

The Visual Cabinet Process Flow

This flow DOES NOT contain time durations at the process, just the flow of product through the system. The only real time that had been established was the three-minute press time at the box press. When the improvement team arrived, the estimated time to complete this production cycle would take anywhere from four to six weeks, sometimes longer.

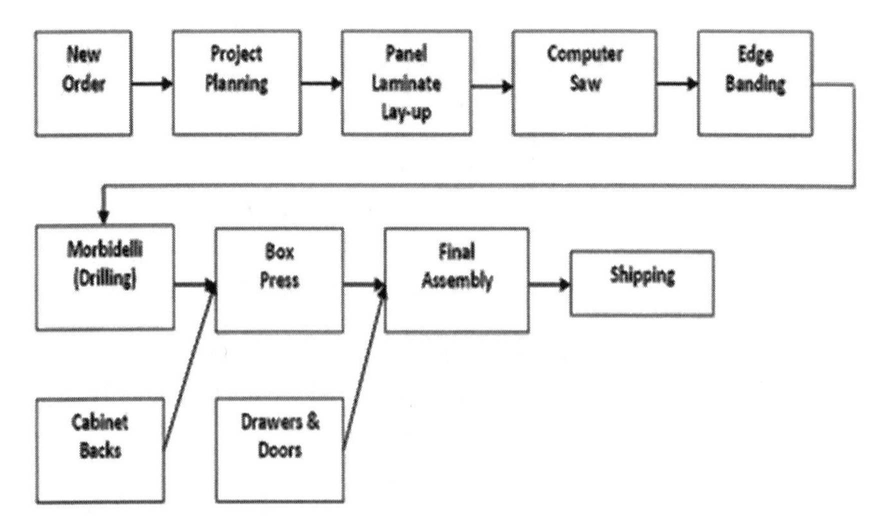

FIGURE 18.1
Visual Cabinet Process Flow.

So, based on your experience and given this much information, you have probably compiled a mental list of ALL the things that need to be fixed or changed. At this point, would you venture a guess as to where the problem area(s) might reside? How come this system takes four to six weeks to produce any results? What, in your estimation, is slowing the system down? How come?

THE SYSTEM ANALYSIS—MORE INFORMATION AND DETAILS

With the tour information in hand, it was time for the improvement team to go back and do a more detailed system analysis and look for the cause-effect-cause relationships at each workstation and determine what the improvement strategy should be. Let's go back through each process step (workstation) and look for more detailed information and clues as to what might be causing the problem.

Step 1—The Planning

The "beloved" optimization software was an add-on from management. Management firmly believed the software could be used to improve efficiency and help them reduce cost-per-part and save on scrap material. In isolation, the software might have some benefits for a single panel, but when subjected to the number of panels required to complete a single project, the software did pose some drawbacks. The foundational logic of the software was to analyze and calculate optimization for a single panel. However, the software did not take into account or calculate the effects on a total project. Overall, the global negative effects of following the software recommendations did cause some problems!

Each project manager usually had several jobs assigned to them to manage. At any given time, the project managers could have three or four jobs at various stages of the planning or production process. The current rules for the project managers were simple:

1. Keep everyone busy (to increase efficiency).
2. Release work to the floor as soon as it was ready to go (in hopes of getting it completed on time).
3. Keep pushing the job through the system.

Step 2—The Lay-up

The lay-up process was pretty straightforward. The capacity of the lay-up team (two individuals) far exceeded the needs of the saw. This workstation was able to keep up or exceed the needs of any of the workstations including doors and drawer fronts—plenty of capacity! The only observed issue for lay-up was sometimes a raw material non-availability. It could happen that extreme laminate colors had been chosen, which required a special order and could delay arrival. These kinds of things didn't happen often, but they could.

Step 3—The Saw

The saw operator was following the wishes of management by trying to extract the highest levels of efficiency that he could from the saw. As such, the saw operator was totally focused on maintaining a very high "efficiency" at the saw because he was measured to make sure the saw was always busy. The internal mantra seemed to be, "If you don't cut it, we can't make it ..." Partially true, but also totally wrong!

Frequently, there were times (daily) the saw operator would "override" the optimization software and change the cutting sequence. Several years of experience had taught him that the software was not always right, and he had developed a better way to do it!

In the saw operator's way of thinking, and also management's way of thinking, this was always a great opportunity to save money. That is, if we could manipulate the process and change things, we could increase efficiency and reduce the cost per part.

In time, the operator had learned that the highest saw efficiency happened when he changed (over-rode) the software to cut the panels differently. Because of his experience, the saw operator was able to look at the entire project cut sequence and determine the number of boxes that were all the same size. As an example, in the case of a school project, it was clear that all of the boxes required in all of the classrooms were basically the same size. In other words, each classroom was the same—almost. Many times, the boxes would vary in size because of a corner unit, or the length of wall, or some other variable, but he could still get very close in his calculations. Besides, he could do the corner units and other various box sizes at a later time. Right now, efficiency was the most important measure to him.

When cutting pieces for the box, the left side and the right side are exactly the same. This could vary based on box height requirements, but even then they were all the same. The tops and bottoms could be a different size depending on the width required and the sequence location (i.e. corner unit), but for the most part the tops and bottoms were the same.

The saw operator reasoned that if he went through the entire job and cut all the left and right sides, before cutting the tops and bottoms, he would be able to maximize the efficiency of the saw by reducing the number of times he had to change the setup and give new instructions to make a new cut. By every measure of efficiency, the saw operator was running at maximum efficiency. The saw operator was very proud of himself for what he could accomplish, and management was also proud of him! He was highly efficient and saving the company a lot of money, or so they thought.

Step 4—Edge Banding

There were two operators at the edge banding machine. One operator would feed the part into the machine, and the other operator would take it out at the other end and send it back for additional edging or put it on the pallet for transfer to the next workstation, if it was finished edging.

For the edge banding operators, it really didn't make much difference which part they had because the process for edge banding was the same. The only thing that really made a difference to them was the color of the edging or if it required wood banding. The banding materials came in big rolls, which made changing from one color to another very easy. Change-out was about two minutes or less.

Step 5—The Morbidelli

The Morbidelli was operated by a single operator. The Morbidelli work deck was actually about 40 inches off of the floor, and the setup of the work deck allowed for two panels to be drilled at a time, sometimes more, depending on the size.

The operator could either drill the flat sides for the pins to hold the shelves or drill the side edges for the dowel pegs. The flat side took a bit more time because there were more holes. The edges took less time because there were only two or three holes for the dowel pegs depending on the size of the piece.

It is interesting to note that the cutting sequence from the saw actually helped the Morbidelli achieve a higher efficiency. The operator could set up to do all of the flat sides first, and then set up to do all the edge holes when the tops and bottoms arrived. So in essence, by following the same cut sequence from the saw, the Morbidelli workstation also appeared to be highly efficient.

Step 6—The Box Press

It was at the box press that the beloved "efficiency" model fell apart! There was at any given time, several pallets of inventory waiting in the holding area for the box press. In some cases, there were pallets stacked on top of pallets (with the help of a forklift), just to better utilize the space available.

There were two box press operators, and they were stuck! Considering all of the work-in-process (WIP) waiting in front of their workstation, they *did not* have enough of the right parts to build a single box! What they usually had was all left sides and the right sides and the backs, but no tops or bottoms. It wasn't until the first tops and bottoms had completed the other processes and started to arrive at the box press that they could start assembly.

When the tops and bottoms did start to arrive, it was sometimes very difficult to match the pieces for the build, the reason being that the width of the boxes had a much higher probability of changing than the height did. At times, it was like putting all the pieces of a puzzle on a table and trying to figure out which pieces went where. Even though most of the boxes were essentially the same size, it usually turned into a "hunt" to find the right pieces for assembly. This problem existed because each piece was marked with job number and cabinet number. Even though each piece was *exactly* the same, the operators would spend considerable time trying to match the job numbers and cabinet numbers!

This matching of the necessary parts also played heavily into the slow production time of the box press. The reality was, the box press operators spent most of their time "looking" for and not "pressing" parts. By all measures, the box press operators were very busy, but they just weren't accomplishing the job of pressing boxes.

Step 7—Final Assembly

The assembly crews in final assembly seemed to have the most down time. I say crews, because there was one crew who did the drawers and another crew who did the doors. Sometimes the hardware could be mounted early for drawers and doors. When the doors arrived, they could have the hinges mounted on them prior to the boxes showing up. However, taking this early action to mount hinges on the doors also generated a large amount of rework. The correct assembly process was to first mount the hinge to the box and then mount the door to the hinge. This allowed them to do the proper alignment when the door was attached.

The door crew often assembled out of sequence, just so they would appear to be busy. They took this action knowing full well they would eventually have to rework it. But, it did meet the measure of being busy and efficient "right now"!

Step 8—Shipping

When the boxes arrived at shipping they were individually wrapped in plastic. They had several different plastic rolls on applicators that they could use and there were, at most times, probably three or four people in the shipping area. They had the capacity to keep up with the boxes coming from final assembly.

When the boxes were wrapped, the box number was checked off the project list and either loaded directly to the truck or placed in a holding area waiting for the truck to return. The primary measure of shipping was to make sure the truck was fully loaded, to maximize the load going to the job site and reduce the number of times the truck had to make a trip, thus reducing the fuel costs and saving money, and of course being more efficient! A truck was rarely allowed to leave for a job site only partly loaded, unless it was to take the last pieces required to hopefully finish the job.

As an added note, there were complaints from the installers at the job sites. Even though lots of boxes were showing up at the job site(s), there were only a few boxes in each room. The installers had some pieces of the puzzle, but not enough of the right pieces to finish the installation.

The System Analysis

With the collection and observation of this additional information, the system started to reveal itself and subsequently take on a life of its own. For the improvement team, the necessary actions started to take shape for what actions were necessary to transform the situation.

By now, many of you might have concluded where you think the constraint is. Based on what you have read (and your mental list of improvements), you've probably deduced which workstation is slowing down the entire system. What do you think the constraint of this system is? Where would you focus your improvement efforts? Why would you pick that particular location or operation?

At first blush, many people might point their finger at the box press. By using traditional system analysis techniques and protocol, (i.e. looking for the slowest operation, looking for the point where the work seems to be backed up and the point that exhibits the highest level of work-in-process inventory), the box press would certainly be a plausible candidate. But, is it really the problem area? If the box press was able to produce at a faster rate, would the entire system get better? Maybe! *Remember*: The box press has its own set of limitations. Each box is required to stay in the press for at least three minutes to set the glue! Can the box press go any faster than it already does with these limitations? How?

THE IMPROVEMENT IMPLEMENTATION STRATEGY

The improvement team made the initial presentation to the management team and shop floor personnel, to share the findings and make the recommendations for moving forward. To sum it up in a few words, the management team and some shop floor workers were "stunned" to hear the recommendations being presented. Remember, up-front the management team had wanted the improvement project focus to be on improving "efficiency" and reducing the "cost per part" even more!

The improvement team recommended that the "efficiency" measure be stopped immediately! The team presented the evidence as to "why" the efficiency measure was causing chaos in the system. Instead of efficiency, the recommendation was presented to move toward "synchronization." The improvement team used the reference environment of a marching band where each member of the band is marching, but they are out of step with those in front of them and behind them.

Both management and some shop floor folks presented arguments about "why" efficiency was so important and necessary to cost savings and cost-per-part reductions. When asked to show "the evidence" of cost savings and reduced cost-per-part, none could be provided. They were all sure the money was there, but nobody could find it. In the end, management conceded to give synchronization a try. (Note: I honestly think that management conceded to give it a try not because they thought it was the right thing to do, but rather because they wanted to prove us wrong!)

Because of the efficiency measure, each workstation was working to a different "drum beat" and therefore totally out of synchronization. By bringing order to chaos, and using synchronization, the improvement team felt like the system could move forward by orders of magnitude.

It's a common situation in many companies where the first constraint is not necessarily on the shop floor, but rather in the policies and procedures used for management. Such was the case for the cabinet makers. The enforcement of the efficiency policy was creating untold consequences and chaos in the system as a whole.

Let's go back through each workstation and present the recommendations from the improvement team.

Step 1—Planners

The basic job of the planners did not change. The architectural drawing still needed to be converted to shop drawings. However, the emphasis was shifted from the entire project to focusing on each room in the project. Cut lists were modified to encompass each room, or building location. The cutting and assembly instruction were now released one room at a time rather than a project at a time. When all the rooms were complete, then the project was complete.

The planners slowed the release of projects to the project managers. Instead of releasing work when they could, they released it only when they should. The WIP dropped dramatically.

Step 2—The Lay-up

Nothing at this workstation changed. The laminate (proper color) was glued to either MDF or plywood panels and moved to the saw. The capacity was sufficient to maintain the new synchronization model.

Step 3—The Saw

Needless to say, the saw operator was "furious" to learn that his efforts at efficiency were meaningless and actually the cause of the system chaos. However, he was on board to give this "synchronization" thing a try.

The rules for the saw operator changed from efficiency to synchronizing the work flow through the system. Instead of cutting the pieces that provided the highest efficiency, the saw operator now cut all of the pieces to provide the parts for a single box. This meant the saw cutting instructions were changed several times for each box. (doors, sides, top and bottom). As the pieces for each box finished cutting, they were placed on a pallet. Each pallet now contained all of the pieces to build a single box, except for the back which still came from a separate location.

Instead of moving a pallet of 30 pieces at a time, the pallet now contained all of the pieces for a single box. Once the system kicked in, there were actually plenty of pallets to do the "one-piece flow" concept. In this case, one piece equals one box. We also assigned a person to move parts from the saw to the edge banding machine. The saw operator was no longer required to move the parts.

Step 4—Edge Banding

For the edge banding machine, the process stayed the same. All they needed to know was edge banding color and type (wood or plastic), and this information was provided in the production package. When the need to change banding color did happen, the setup could be completed in two minutes or less.

In essence, they now banded all of the parts for a complete box (sides, top and bottom, doors and drawer fronts). When the parts finished, they were placed back on the pallet and moved to the Morbidelli. The same material mover assigned to move parts from the saw to edge banding also moved the pallets to the Morbidelli before going back to the saw for the next pallet.

Step 5—The Morbidelli

The Morbidelli operator was also annoyed by the shift from efficiency to synchronization. He too had more computer commands to be entered based on the parts being drilled (sides, dowel holes and doors). However, the transition was smooth and presented no real problems.

Step 6—Kitting

In the previous system, parts moved from the Mobidelli to the box press. In the new system, the Kitting step was added to prepare everything prior to the box press. We discovered the need for this step by using an Interference Diagram. At the center of the diagram we concluded "More box press time" and determined what the interferences were that stopped them from getting that. One of the major interferences was that the press operators were spending too much time looking for parts. The injection (idea) was for the operators to spend less time looking for parts, and subsequently we off-loaded that task to a new process step we called "Kitting."

At the Kitting step, the boxes where checked to make sure ALL of the parts were accounted for compared to the project checklist. If something was missing, or out of place, the job was placed in a "hold" area until the missing parts could be found. The Kitting team also moved the doors and drawer fronts to the final assembly area. Doors and drawer fronts had no need to go through the box press. The box backs were added to the pallet to make a complete kit and moved to the box press.

Step 7—The Box Press

For the box press team, life was wonderful! Each pallet that arrived had all of the necessary pieces to press a complete box. As one box was being pressed (three minutes), the team did the rough assembly on the next box. When the box in the press finished, the pressure was released, and the box moved out into the receiving area for final assembly. The rough assembled box was moved into the press, the pressure applied, and the next box was rough assembled.

Step 8—Final Assembly

The rework and confusion for final assembly dropped dramatically. The team no longer felt the pressure to assemble a door ahead of time just to look busy. The doors and drawer fronts for the box coming out of the press had already arrived a few minutes prior. Now the proper assembly sequence could be conducted by first hanging the hinges on the box and then attaching the doors. Door and drawer hardware (pulls) were added and the box moved to shipping.

Step 9—Shipping

Although the actual process steps for shipping didn't change much, the boxes they were getting took on a new meaning. Shipping now understood they were receiving boxes based on a specific sequence. The new sequence of receiving boxes based on a particular room, and not just a job site, helped in the loading sequence of the truck.

To reduce the build-up of WIP in the shipping area, a new policy for gaining truck capacity had been implemented. Instead of buying any new trucks to increase the capacity, they instead rented the trucks as necessary, to move product to the job site. Even though it was an added operational expense to rent trucks, it was also much less expensive than buying a new truck. The Throughput gained from the extra truck capacity far outweighed the expense.

Step 10—Installation

The installers were a much happier group! The change from efficiency to synchronization allowed them to receive all of the necessary boxes to complete the installation in a single room. No longer did they have several rooms almost done and waiting for more boxes, but now they could complete a room and move to the next one.

The Results

As simple as all of this sounds, the improvement approach was not readily apparent to the management team. Their focus, which they had for years, was in the wrong direction. The primary focus of trying to maintain high efficiency, reduce the cost-per-part and save money had not served them well.

The real improvement effort turned out to be exactly the opposite of what they were thinking. Instead of a high energy focus on efficiency—forget about it! Instead of trying to reduce the cost-per-part through efficiency—forget about it! Instead of focusing on trying to save money, shift now to the focus of "Making Money." The strategy to save money is much different than the strategy to make money.

The lead time through the shop, from the time wood started lay-up until it ended up on the shipping dock, was reduced from four to six weeks to only four hours!!

There is good news and bad news about the final, end results. First, because of the improvements implemented and the results achieved, the competitors took notice. This cabinet maker was eventually bought out by one of the competitor companies at a very attractive selling price. The bad news is that, shortly after acquiring this company, the new owners reverted back to standard Cost Accounting methods, including efficiency, reduced cost-per-part and cost savings. They filed for bankruptcy and went out of business about 14 months later.

Index